A GUIDE TO FLUIDICS

A Guide to Fluidics

edited by

ARTHUR CONWAY

MACDONALD : LONDON
AMERICAN ELSEVIER : NEW YORK

Sole distributors for the United States and Dependencies
American Elsevier Publishing Company, Inc.
52 Vanderbilt Avenue
New York, N.Y. 10017

Sole distributors for the British Isles and Commonwealth
Macdonald & Co. (Publishers) Ltd
P.O. Box 2 L.G.
49–50 Poland Street
London W.1

All remaining areas
Elsevier Publishing Company
P.O. Box 211
Jan van Galenstraat 335
Amsterdam
The Netherlands

Library of Congress Catalog Card Number 71–141516

British SBN 356 02799 6
American SBN 444 19601 3

Made and printed in Great Britain
by Butler & Tanner Ltd, Frome and London

CONTENTS

ACKNOWLEDGEMENT

Dr Keith Foster, of the University of Birmingham, gave his advice to the editor freely and generously during the early and formative stages in the preparation of this book. The editor is glad to acknowledge his debt and to express his gratitude to Dr Foster for the help received.

1

FUNDAMENTALS OF FLUIDICS

PROFESSOR E. MARKLAND
University College, Cardiff

DR R. F. BOUCHER
University of Sheffield

For many years engineers have used fluids for transmitting force and energy in mechanical systems. For example, the flexibility and great power offered by hydraulic actuators and motors make them attractive for use in many control systems, even when the systems are essentially electrical or electronic. The predominance of electronic techniques in computation and in control systems may give the impression that such systems can function only electronically. In fact, Babbage's calculating 'engine', which was perhaps the earliest digital computer, with the essential conceptual features of present-day machines, was purely mechanical, and various simple, completely pneumatic, control systems, such as automatic pressure controls, have been in commercial use for many years. It is only comparatively recently, however, that techniques have been developed for using fluid as a medium for computation and for carrying significant information in control systems.

Hydraulic and pneumatic circuits which use valves have been developed for many automatic processes, for example those requiring a controlled sequence of events. The conventional valves used in such circuits have mechanical moving parts and may be classed as *static* devices. The term arises because forces are developed by the static pressure of the fluid, power being expended only intermittently when the fluid is in motion.

A new class of fluid circuit element has now been added which may be referred to as *dynamic*. In dynamic elements the kinetic energy of the flowing fluid is all-important, and in them power is expended continuously. An important aspect of such elements is that they need not depend on moving parts for their operation, so there is promise of long life and high reliability.

1.1 The birth of fluidics

The inauguration of the new technology of fluidics, which has developed from the invention of dynamic elements, is usually attributed to work announced in 1960 at the Harry Diamond Laboratories in Washington. Since that date there have been three Harry Diamond Conferences (*1*) and four Cranfield Conferences (*2*) as well as other meetings at which developments have been reported. Various estimates of the future growth of the fluidics industry have been given during this period, and it is not surprising that opinions vary widely as to promising lines of application and the ultimate position of the technology in relation to the whole of control engineering. A survey completed in 1969 (*3*) estimated that, up to the end of 1967, total expenditure on fluidics in the U.K. had been £2·5m, and market estimates of £3m per year in 1972 and £5m to £15m per year in 1975 were advanced: it was thought that the European market would probably be three times greater in 1975.

It should be added that the invention of elements having no moving parts has generated a new interest in control valves *with* moving parts. Diaphragm and ball-operated elements for control circuits have been significantly developed during the past ten years. Although these are static elements according to our terminology, they are usually included under the general adjective *fluidic*, while the term *flueric* is sometimes reserved for elements of the dynamic type.

Certain technical advantages have frequently been claimed for fluidics: simplicity leading to high reliability, ability to operate in extreme conditions of temperature, resistance to shock and vibration, safe operation in explosive atmospheres, freedom from interference from radiation, and so on. Additionally, the cost of the elements is potentially low, although it is by no means certain that the market will ever be sufficiently large to make this advantage decisive in many applications where alternative means are convenient.

1.2 Fluid mechanics of a jet

The fluid jet appears in one form or another in most dynamic elements, so it serves as a useful example for describing much of the relevant fluid mechanics. Intuitively one would expect a fluid jet emerging from a nozzle into a large volume of fluid to behave as indicated in fig. 1.1. The velocity distribution across the nozzle exit is uniform, or very nearly so, and the jet interacts with the fluid into which it moves.

The property of a fluid known as its *viscosity* is its resistance to relative motion between adjacent layers. A viscous shear stress decelerates the outer layers of the jet while accelerating the adjacent layers of the stationary body of the fluid. Therefore velocity is redistributed as in fig. 1.1. A central core, moving almost with the original speed, is enclosed by shear layers across which the velocity declines smoothly to zero. The width of these layers increases as an

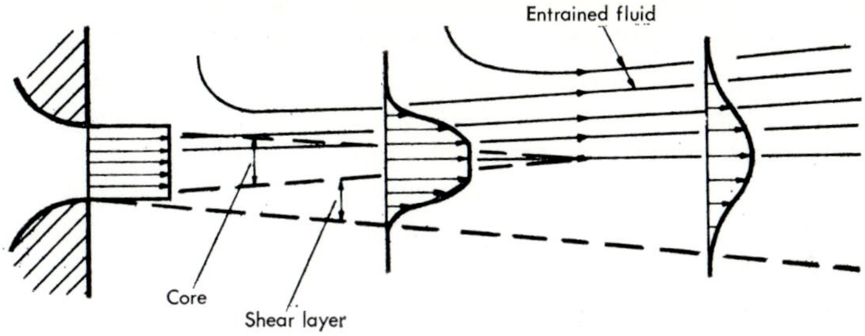

Figure 1.1 Laminar jet.

increasing amount of fluid is entrained by the jet, spreading the jet and narrowing the central core, until the layers meet in the centre. Subsequently the centre-line velocity decreases asymptotically to zero, the original kinetic energy being dissipated by the viscous action. Momentum, however, is conserved along the length of the jet, the reduction in mean velocity being exactly compensated by an increase in the mass flow caused by entrainment.

A jet answering the above description is called *laminar*. The motion is completely steady and shear stress between adjacent layers in

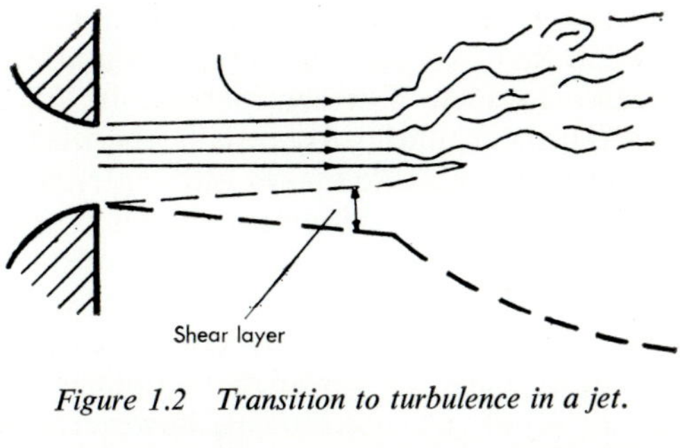

Figure 1.2 Transition to turbulence in a jet.

relative motion is due to the viscosity of the fluid. In practice, however, the flow in such a jet breaks down and becomes turbulent at some position along its length; the phenomenon may often be seen in the smoke of a cigarette.

Turbulent motion, or random disturbance superposed on a mean flow, greatly speeds the spread of a jet and reduces its penetration into a body of fluid. This is indicated in fig. 1.2. If conditions are such that the jet remains laminar for a great length, say a hundred nozzle widths, then transition to turbulence may be observed as a growth of periodic disturbances which start as waves in the jet. When the laminar length is short, however, this mechanism is not apparent, and transition from laminar to turbulent flow is direct, as shown.

The mechanism of transition is not understood, but the position along the jet at which transition occurs is determined by the size and speed of the jet and by the properties of the fluid. The quantity which characterizes the flow in terms of these quantities is the Reynolds number, *Re*, which is defined as—

$$Re = \frac{\rho U d}{\mu} \tag{1}$$

in which ρ denotes the density and μ the coefficient of viscosity of the fluid, U denotes the speed, and d denotes a characteristic dimension of the system, which in this case could conveniently be the width or diameter of the nozzle. For dry air at 15°C and 1 atm pressure—

$$\rho = 1{\cdot}226 \text{ kg/m}^3 = 0{\cdot}0765 \text{ lb/ft}^3$$
$$\mu = 1{\cdot}79 \times 10^{-5} \text{ kg/m s} = 1{\cdot}20 \times 10^{-5} \text{ lb/ft s}$$

The ratio of these quantities is usually called the coefficient of dynamic viscosity, ν, which for air has the value—

$$\mu/\rho = \nu = 1{\cdot}46 \times 10^{-5} \text{ m}^2\text{/s} = 1{\cdot}57 \times 10^{-4} \text{ ft}^2\text{/s}$$

Experiments show that jets useful for fluidic applications occur with *Re* between about 400 and 20 000, and that, in this range, the laminar length decreases as *Re* increases. For a circular-section jet, issuing into reasonably still surroundings, the laminar length may be 300 diameters when $Re = 500$, and this falls to about eight diameters when *Re* increases to 1500. Further increase in *Re* shortens the laminar length even more, until substantially the whole of the jet is turbulent. Further increase in *Re* makes little difference to the flow.

In the range in which *Re* is significant the jet is sensitive to disturbance in the surroundings. For instance, acoustic disturbances were shown by Lord Kelvin to shorten the laminar lengths of gas flames. This sensitivity of a long laminar jet to disturbance is used in an important fluidic element, the turbulence amplifier. In the range in which virtually the whole jet is turbulent, however, the jet is comparatively unaffected by disturbance in the surroundings.

Yet a disturbance in the flow upstream of the nozzle, and the shape of the nozzle, both affect the rate of spread of the jet. For a two-dimensional turbulent jet the central core disappears about five nozzle-widths downstream of the nozzle, and the subsequent angle of spread is about 15° to each side of the centre-line. This is much greater than the spread of a laminar jet: turbulent fluctuations are much more effective in entraining flow from the surroundings, and in transferring momentum from one streamline of mean flow to another, than is viscous shear in a laminar jet.

1.3 Jet reattachment

A solid wall adjacent to a two-dimensional jet restricts replacement of fluid entrained by the jet from the space between itself and the wall, as shown in fig. 1.3(a). Thus the pressure in the space falls in relation to that on the other side of the jet, bending it towards the wall. The restriction is thereby intensified and the deflexion increased,

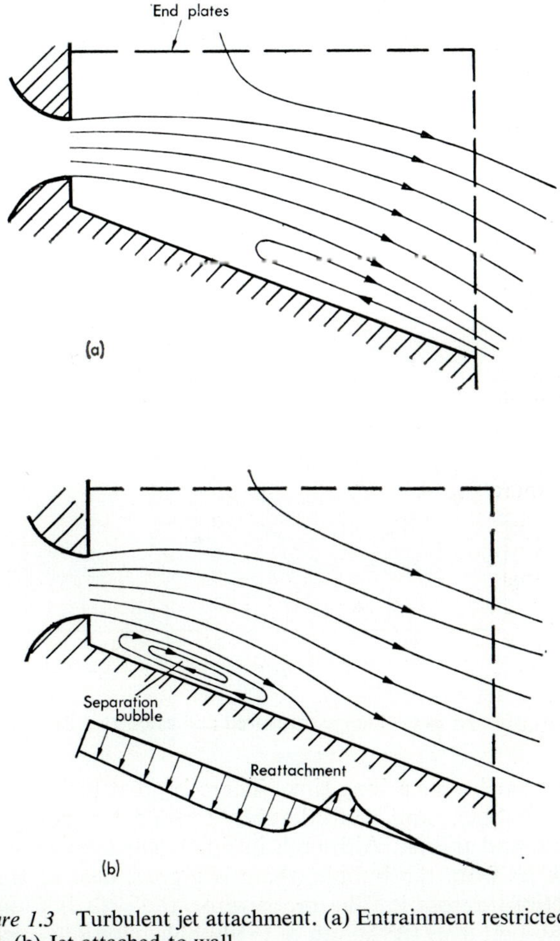

Figure 1.3 Turbulent jet attachment. (a) Entrainment restricted by wall. (b) Jet attached to wall.

until ultimately the jet attaches itself to the wall in the stable configuration shown in fig. 1.3(b). There is then a 'separation' bubble between the jet and the wall, closed at the downstream end where the jet is reattached. Fluid is entrained from this bubble by the jet, and the bubble is replenished with fluid from the reattachment zone. The flow lines shown in fig. 1.3 are streamlines of mean flow, i.e. lines of

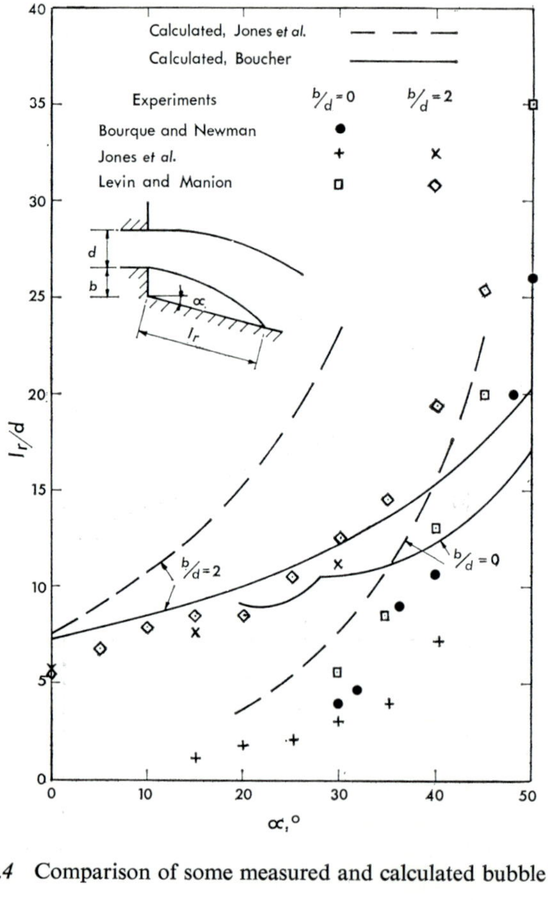

Figure 1.4 Comparison of some measured and calculated bubble lengths.

motion averaged over a long time. There is great turbulence in the separation bubble, and considerable turbulent mixing between bubble fluid and the jet. Although in fig. 1.3(b) a mean streamline divides the jet from the bubble, there is a great deal of transfer in both directions across it. The pressure on the wall is found to be roughly constant over the length of the bubble, rising to a maximum slightly downstream of the reattachment point (where the part of the jet that returns divides from that which continues downstream) and finally falling towards the pressure in the undisturbed fluid.

In spite of many attempts, no completely satisfactory method of analysis has been evolved. The early treatments by Bourque and Newman (*4*) and Sawyer (*5*) assumed simple models of the jet and constant pressure in the separation bubble to predict the bubble length. These were followed by various refinements. Attempts were made to allow for the effect of curvature on the rate of entrainment into the jet (*6*). Jones *et al.* (*7*) tried to calculate the variation of pressure along the length of the bubble and the effect on bubble length. Boucher (*8*) sought what degree of improvement a more realistic model of the jet might afford. Some of the calculated results are compared with measurements in fig. 1.4.

These analyses are based on fully-turbulent jets. As the value of *Re* falls below, say, 5000, the bubble tends to lengthen while the laminar length of the jet begins to increase. If the wall is of finite length the bubble will ultimately extend to the end as *Re* is successively decreased, and the jet then detaches itself. A further complication arises for jets of finite depth. Figure 1.3 indicates the jet to be contained between end plates; the ratio of jet depth to nozzle width is referred to as the *aspect ratio* of the nozzle. Boundary layers form at end plates, and these make leakage paths from the region above the jet into the low-pressure bubble region below. The leakage lengthens the bubble to an extent which is usually significant when the aspect ratio is less than about 2.

The effects of *Re* and aspect ratio are important in practice because it is usually desirable to make elements as small in size and power consumption as possible. Many commercial elements have nozzle widths of less than 0·5 mm, operating in the range of *Re* below 5000. For any chosen nozzle width and supply pressure the power consumption is proportional to aspect ratio, so is least with the smallest aspect ratio giving satisfactory jet attachment.

1.4 The beam deflexion amplifier

The most simple element is the *beam deflexion amplifier* (cf. Section 2.4.1). In it a jet flows from a rectangular nozzle towards two output passages which may be separated by a simple dividing edge or splitter, or which may have a centre vent between them. The flow is usually confined between parallel plates so that the jet spreads only in one plane as indicated in fig. 1.5. Control nozzles are provided on each side of the jet so that if a differential pressure is applied across them the jet is deflected towards one or other output channel.

Deflexion may be caused simply by momentum interaction, or, if the interaction chamber is suitably designed, by pressure effects on the sides of the jet.

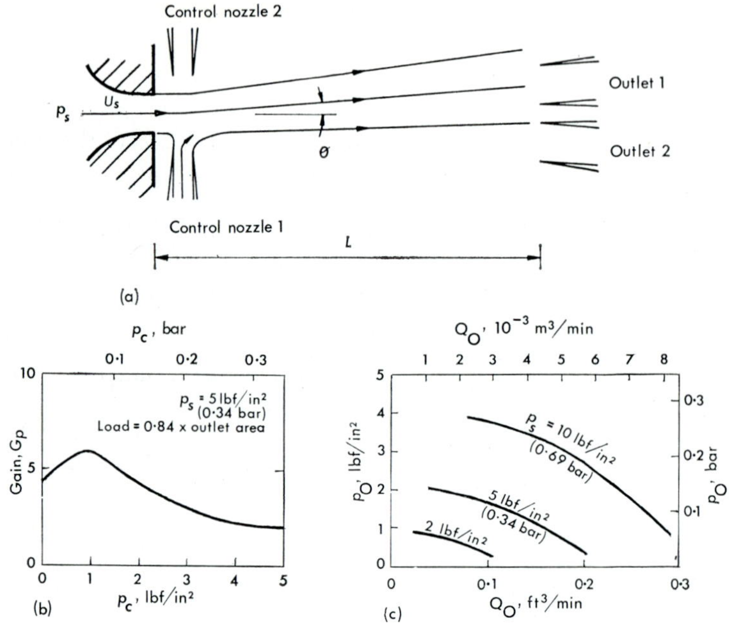

Figure 1.5 Free-jet amplifier. (a) Schematic diagram. (b) Variation of pressure gain. (c) Output characteristic.

The deflexion angle θ of a jet possessing momentum J_s, deflected by a control jet possessing momentum J_c at right angles to it, is given by—

$$\tan \theta = J_c/J_s \tag{2}$$

Were it possible to place the output channels very far downstream, a very sensitive amplifier could be produced. However, as we have seen, the energy of the jet is dissipated by mixing, so the output receivers are positioned far enough away to require only moderate angular deflexion, but close enough to recover a reasonable proportion of the kinetic energy of the jet.

The element normally operates as an amplifier of the difference between signals across the control ports, and it is convenient to use the pressure gain, G_p, defined as—

$$G_p = \frac{dp_o}{dp_c} \tag{3}$$

where the subscript *o* denotes outlet and *c* denotes control, and the outlets are either blocked or loaded in a prescribed manner. There is a corresponding flow gain—

$$G_f = \frac{dQ_o}{dQ_c} \tag{4}$$

Some typical gain and output characteristics are shown in fig. 1.5. The gain depends on the bias pressure applied at the control nozzles. Note that the maximum output pressure is about 40% of the supply pressure at the main nozzle; this is typical of many proportional and digital fluidic amplifiers.

The speed of response is also often important. If U_s is the velocity and L the length of the jet, then the delay between a control signal and the appearance of a disturbance at the output is approximately—

$$\tau = L/U_s \tag{5}$$

The turbulence in the jet causes noise at the output. It has been shown that, by reducing the aspect ratio, the signal/noise ratio may be improved (*9*). By providing a number of parallel paths, each of low aspect ratio, values as high as 1500 have been obtained for the signal/noise ratio. By using a laminar jet the problem of noise can be eliminated, but it is difficult to deflect a laminar jet without making it turbulent. Some progress has been made in this direction (*10*), but laminar jet elements are not yet available.

1.5 The *and* and *exclusive or* element

As a further example of a fluidic element which depends, like the free-jet amplifier, on momentum interaction, consider the *and* element.

Figure 1.6 shows two control nozzles, C_1 and C_2, placed at right angles to each other so that, with signals at both nozzles, the flows combine to produce a single jet which gives an output at the centre outlet, O_2. This is the function of an *and* gate, output O_2 being present only if both C_1 and C_2 are present. In the presence of only C_1 or C_2 there is an output at one or other of the outlets O_1 and O_3.

This element may be classified separately from the free-jet amplifier on two counts. Firstly there is no power jet to be deflected, interaction occurring between control jets only. Such an element may be referred to as *passive* where an element using a continuous power jet is called *active*. Secondly, the element is digital—signals are either at the ports or not—and there are no intermediate signal levels.

A typical silhouette and some characteristics are given in fig. 1.6. The maximum output pressure at the *and* port, O_2, is about 80% of the control pressure when the output is blocked, and this falls to about 60% when the outlet flow rate is equal to the flow rate in either of the controls.

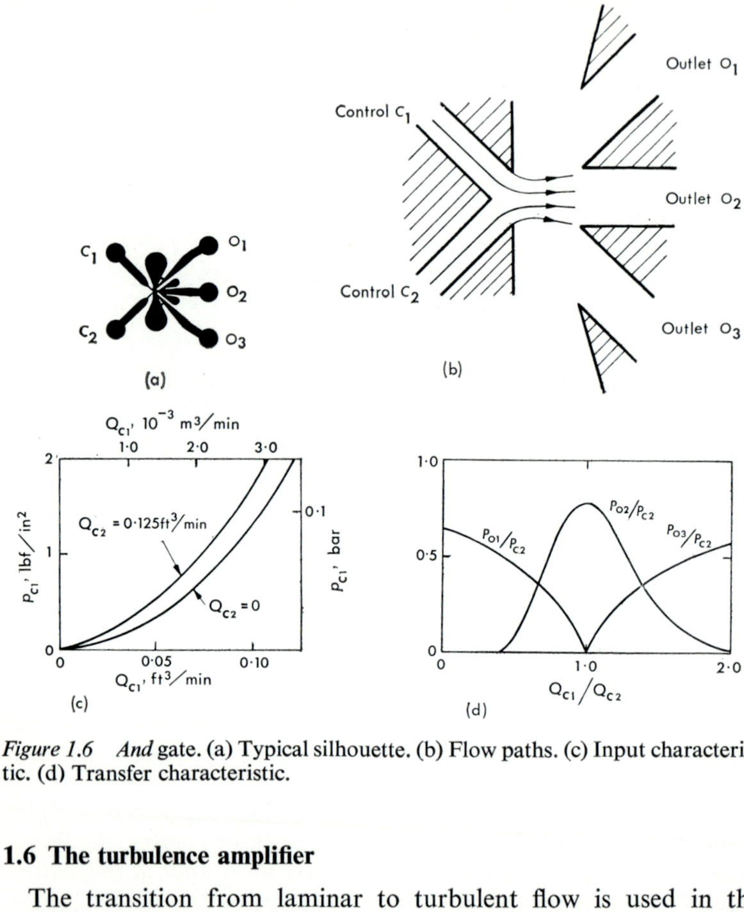

Figure 1.6 *And* gate. (a) Typical silhouette. (b) Flow paths. (c) Input characteristic. (d) Transfer characteristic.

1.6 The turbulence amplifier

The transition from laminar to turbulent flow is used in the *turbulence amplifier*, an active digital device which performs the *nor* function (cf. Section 2.1).

A laminar jet is produced, as shown in fig. 1.7, in a long parallel tube or duct which tends to remove disturbances originating upstream and forms a rounded velocity profile at the start of the jet. The Reynolds number of the flow is chosen so that, in the absence of control signals at any of the control nozzles, the jet remains laminar over the length to the output tube. When pressure is applied at any one or more of the control ports, the jet is disturbed and becomes turbulent before reaching the outlet channel. Because diffusion in the turbulent jet is much more rapid the output signal falls from its previous value to a much smaller one.

The elements are usually manufactured in planar form with the space between the supply and outlet channels vented to atmosphere (for details of construction see Chapter 2). This conveniently ensures that there is no cross-talk between the control ports. The number of

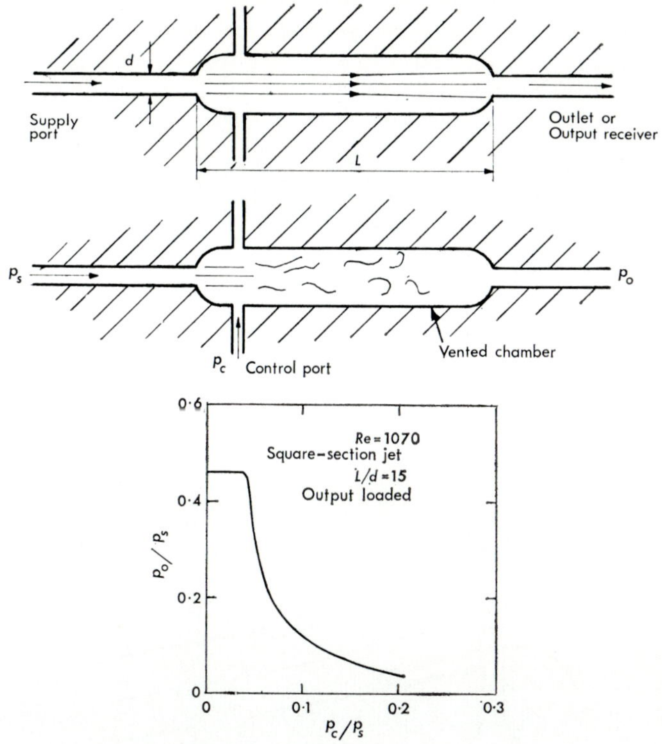

Figure 1.7 Turbulence amplifier—principle and characteristic.

control ports may be large, although it may be standardized at two or four for economy in production. The *fan-out* ratio, i.e. the number of elements which may be switched by the output of a single element of the same type, is typically about four (*11*).

The jet does not become laminar again immediately on removal of the control signal, probably because of the feedback of turbulent flow from the region of the outlet channel towards the control ports. The delay is typically about 4τ, where τ is the time for transport of a particle along the length of the jet. Various methods have been suggested for reducing this, and at the same time improving the difference between the levels of output signals when *on* and when *off*. The methods usually involve an edge or wall, close to the jet at some position along it, to promote turbulence very rapidly once laminar flow breaks down, or to cause turbulent attachment and deflexion of the jet.

1.7 Wall reattachment (or 'Wall attachment') amplifiers

Most of the research work has been on those fluidic elements which depend on wall reattachment, as already described (Section 1.3), since they have considerable promise as digital amplifiers. If the

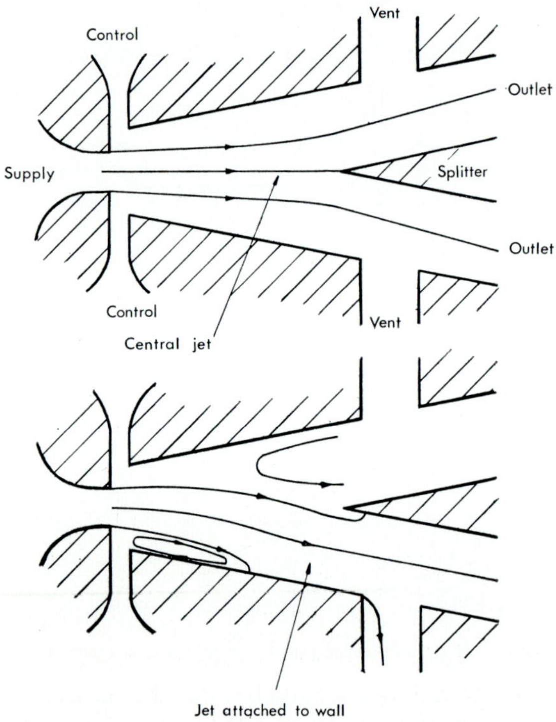

Figure 1.8 Wall reattachment amplifier.

supply of fluid to the nozzle of the completely symmetrical amplifier shown in fig. 1.8 is gradually increased, then at small flow rates the jet will be laminar, and it will flow down the centre line, dividing at the splitter. As the speed is increased the jet will attach itself to one wall or the other, and this condition is identified by the upper minimum Reynolds number, Re_u. If the speed is now slowly reduced, the jet will remain attached for a certain range until the lower minimum Reynolds number, Re_l, is reached, when it spontaneously detaches itself and reverts to the centre-stable flow pattern.

The normal range of operation of the element is above Re_u. The jet remains attached to one wall until a control signal of sufficient intensity and duration arrives at the control port on the attached side. Control fluid flows into the separation bubble, disturbing the balance between entrainment and recirculation, making the bubble grow. Switching then follows by one of two mechanisms. The bubble may lengthen until it becomes detached at the downstream end and the jet moves across to the other wall. Alternatively, the jet may be deflected sufficiently for attachment to the other wall near the nozzle, while remaining for a time attached to the first wall further downstream: switching is completed when the jet moves across the downstream end of the interaction region. The element is therefore bistable, operating as a flip-flop.

If the pressure at the supply nozzle continues to increase sufficiently, the element may again become centre-stable, not owing to Reynolds number effects but because of other fluid flow phenomena.

If the fluid is a gas the effect of its compressibility can be to detach the jet when the speed of the gas is sufficiently high. The influence of compressibility is indicated by the *Mach number*, denoted by *Ma*, which expresses the flow velocity U_s in terms of the propagation speed of an acoustic wave, *a*, as—

$$Ma = U_s/a \tag{6}$$

If the fluid is a liquid the pressure in the bubble may fall to the vapour pressure of the liquid as speed increases. Further increase of speed will then lengthen the bubble since the pressure in it can fall no further, and detachment of the jet will follow. The usual measure of this effect is in terms of the *cavitation number*, σ, defined by—

$$\frac{p_a - p_v}{\frac{1}{2}\rho U_s^2} \tag{7}$$

in which p_a is the pressure outside the bubble, p_v the vapour pressure of the liquid, and $\frac{1}{2}\rho U_s^2$ the dynamic or velocity pressure in the jet. The value of p_v for any liquid varies with the temperature, as indicated by the following values for water.

Temp. (°C)	*Vapour pressure (lbf/in²)* *Absolute*	*Gauge*
15	0·25	−14·45
50	1·79	−12·91
100	14·70	0·00

For gases it is the Mach number, and for liquids the cavitation number, which places an upper limit on the supply pressure under which bistable action will occur in a particular shape of element.

There is frequently a hysteresis zone at the upper end similar to that at the lower end. Over the hysteresis zones the central jet is stable as well as the attached flow configuration, as indicated in fig. 1.9. Operation in a tristable mode is made possible by the zones,

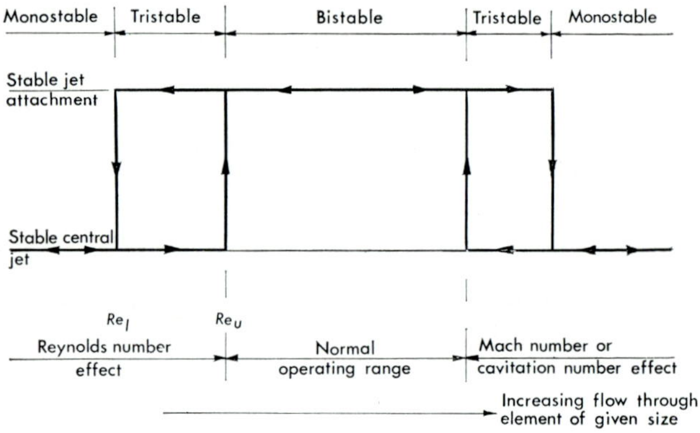

Figure 1.9 Effect on wall-attachment amplifier of varying flow rate over wide range.

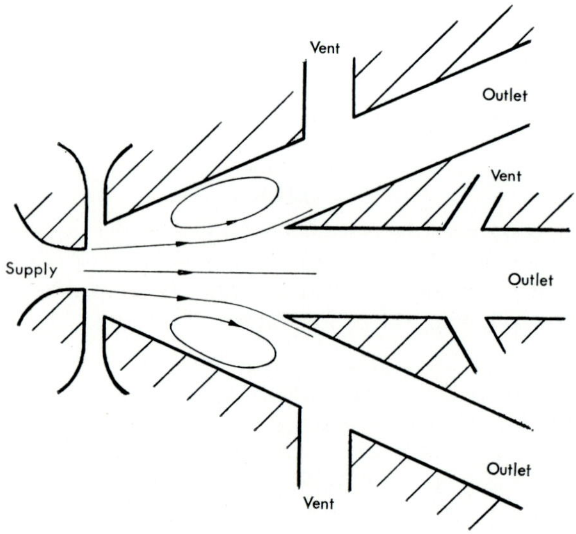

Figure 1.10 Tristable amplifier with three outputs.

a signal at either control port causing attachment to the opposite wall. Simultaneous signals at both ports restore centre-stable flow. It is also possible to operate below Re_l so that the jet is stable only in the centre, but can be deflected to either wall by one or any other control signal. In the absence of any control signal (as well as in the presence of both simultaneously) the jet will always return to the central position. The typical configuration of an element designed to operate in tristable or centre-stable modes is indicated in fig. 1.10. This is a logically powerful unit, but no version is available commercially at the time of writing.

The aspect ratio also has an important influence on the minimum Reynolds number. Glaettli, Müller and Zingg (*12*) have made measurements, reproduced in fig. 1.11, on the upper and lower

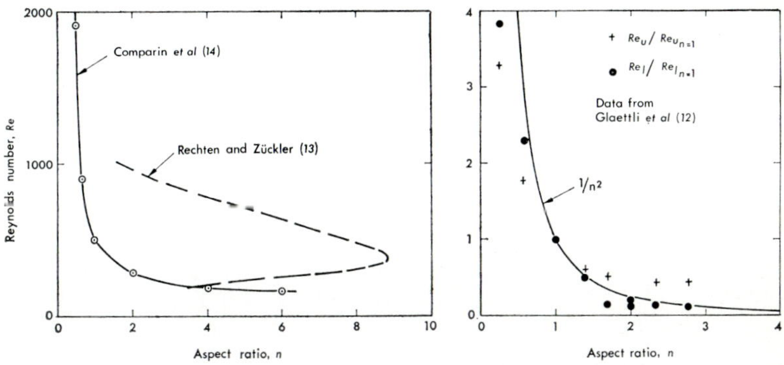

Figure 1.11 Effect of aspect ratio on minimum Reynolds number.

critical Reynolds numbers as functions of aspect ratio. Because of the sharp increase in Re_u and Re_l as the aspect ratio n falls below about 2, most elements are manufactured with n not less than this value. Figure 1.11 also shows measurements by Rechten and Zückler (*13*) and by Comparin *et al.* (*14*) indicating that values of Re may be kept low in certain carefully designed elements.

The power consumption P of an element which uses incompressible fluid of density ρ supplied at pressure p_s is given approximately by—

$$P = nd^2\sqrt{(2p_s^3/\rho)} = \frac{\mu^3}{2\rho^2}\frac{n}{d}R_e^3$$

For elements of given geometrical proportions the power therefore increases as the square of the size if the supply pressure is held constant, but *increases* as the size *decreases* if the Reynolds number is held constant (the pressure required to keep it constant varying inversely as the square of the size). Now it is usually desired to make the elements as small as practicable to minimize switching time. Thus, to maintain the necessary value of *Re*, the power consumption must rise as the size of the element is reduced. The switching time of a given element depends on the pressure applied at the control port; if the applied pressure is substantially in excess of the minimum required to achieve switching it is typically 10τ, where τ is the transport time of a particle moving with the initial jet speed over the length of the reattachment wall. If near-minimum pressures are applied, the switching time can become three or four times as large and can have the large random variations sometimes referred to as *jitter*.

Considerable work has been put into numerous detailed studies of the effects the shape of an element has on its performance. The wall length must be chosen in relation to the wall angle and offset for

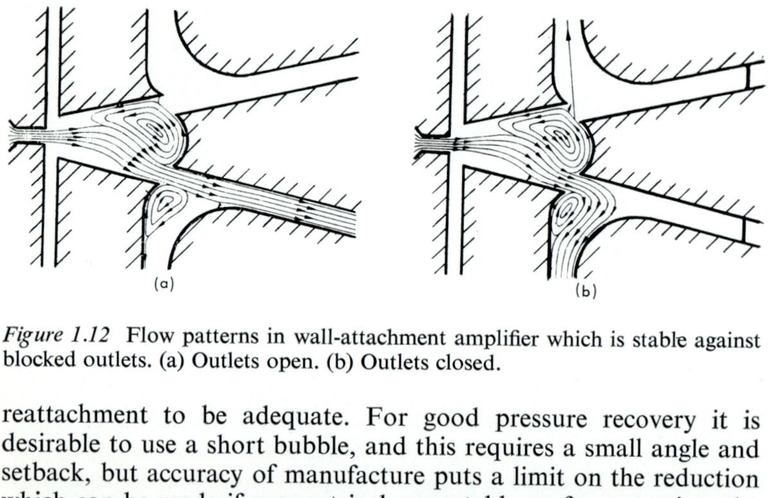

Figure 1.12 Flow patterns in wall-attachment amplifier which is stable against blocked outlets. (a) Outlets open. (b) Outlets closed.

reattachment to be adequate. For good pressure recovery it is desirable to use a short bubble, and this requires a small angle and setback, but accuracy of manufacture puts a limit on the reduction which can be made if symmetrical, repeatable performance is to be obtained. The splitter is crucial in determining performance with loaded output. A simple sharp-edged splitter will merely divide the flow as the load on the active output passage is increased, causing spurious flow into the other leg. Flat-ended and cusped splitters are often used to deflect any excess flow back into the interaction region. A reverse vortex may be formed on the outside of the jet, strongly stabilizing the jet attachment. Similar effects may be found by shaping the vents, and fig. 1.12, taken from Rechten (*15*), shows a configuration which is stable under blocked load conditions and which gives good pressure recovery.

Figure 1.13 shows a typical silhouette and characteristics of an element having a nozzle 0·020 in (0·51 mm) wide. The maximum output pressure, under blocked load conditions, is about 45% of the supply pressure p_s (3 lbf/in^2). The switching characteristic shows hysteresis. With slowly increasing control flow, a pressure of $0{\cdot}05p_s$

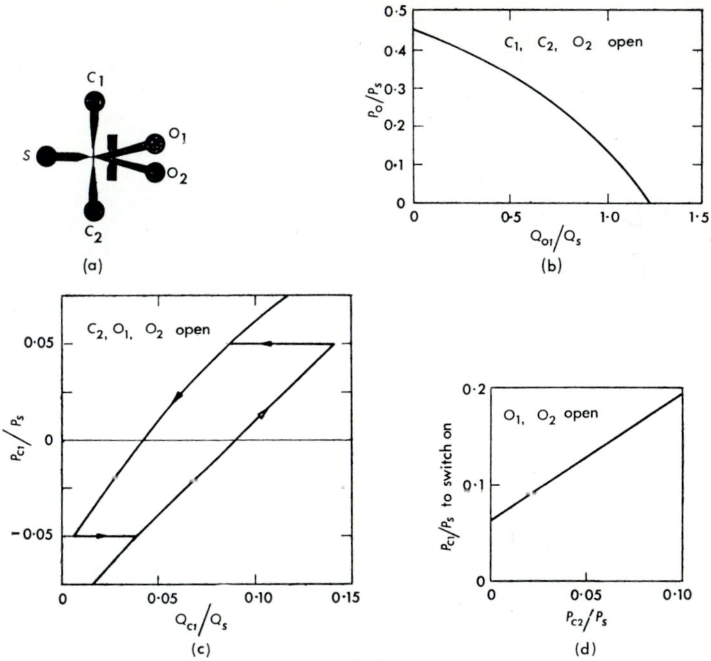

Figure 1.13 Bistable. (a) Typical silhouette. (b) Characteristic. (c) Switching characteristic. (d) Switching characteristic with bias flow.

is required to detach the jet, at which point the control pressure decreases suddenly because of the change in pressure when the bubble disappears. Restriction of flow to the control nozzle causes reverse switching at a lower pressure.

1.8 The vortex amplifier

The *vortex amplifier*, illustrated in fig. 1.14, is a flow-modulating element (cf. Section 2.4.2).

The flow is contained in a cylindrical chamber supplied radially through nozzles or a permeable surface at the perimeter, and leaves by a hole or holes in the centre of one or both of the end plates. In the absence of a control signal the flow is radial from the outer surface to the centre. The flow is modulated by means of a control flow through one or more tangential nozzles round the perimeter.

This imparts swirl to the motion, generating a radial pressure gradient positive from the centre. The flow resistance of the chamber is therefore increased. The details of the swirling motion are extremely complex. The flow in the boundary layers near the end

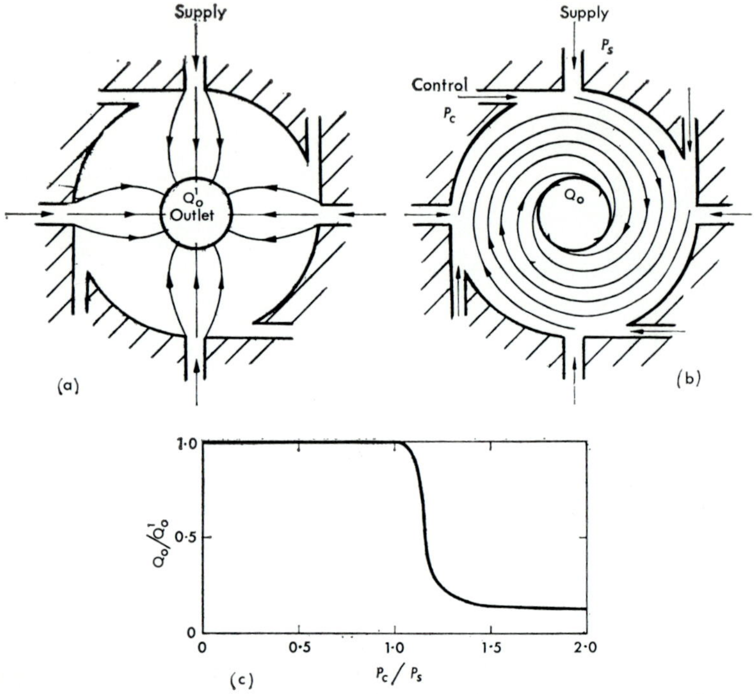

Figure 1.14 Vortex valve. (a) Radial flow without control signal. (b) Swirling flow produced by control signal. (c) Control characteristic.

surfaces of the cylinder rotates less rapidly than the flow at mid-height, and the radial component of velocity may reverse over the central portion of the height. The modulation ratio, or turn-down ratio, may be as high as 25 : 1, although there may be trouble with undesirable oscillation if it is made too high (*16*).

The vortex chamber may be adapted in many convenient ways. It may be operated differentially by using opposing control tubes. It becomes a fluidic 'diode' if the radial supply flow is omitted; the forward flow direction of the diode is from the centre outwards (cf. Section 5.2). A vortex chamber without control tubes, rotated about its axis, makes an angular velocity sensor: the tangential velocity at the centre is a sensitive measure of the rate of rotation (see also Section 6.2 and Chapter 7, especially Section 7.4).

1.9 Elements with moving parts

No introduction to fluidics would be complete without mention of the development of elements from control valves of a more conventional nature. The free-foil element (*17*) is, for example, an adaptation of the disc non-return valve which has been well established over many years for air compressors. The conventional spool valve has been developed for pneumatic control circuit work, and is increasingly being thought of as a logic element, under the stimulus given by the development of pure fluid elements. There is also the possibility of making truly fluidic elements, such as wall-reattachment amplifiers, with slightly deformable parts. Performance and utility in some applications may thereby be substantially improved without reducing life and reliability to any significant extent.

Of the many elements which have been proposed we shall describe only two, both of which are commercially available in the United Kingdom.

The Techne spring element is essentially a *nor* gate. Air admitted at the supply connexion, shown in fig. 1.15, passes through the restrictor to the interior of a close-coiled helical spring, and gives a signal at the output. A signal at much lower pressure to either of the control connexions causes the flexible diaphragm to bend the spring, allowing leakage to the atmosphere through the opened coils so that the output pressure falls to a small value. The units are manufactured with two springs per diaphragm, as the photograph shows. The characteristics are similar to those of a turbulence amplifier, but the operating pressure is much higher and the response time is longer. It is worth noting that power is consumed only while there is a control signal, and that in some applications it is possible to operate the control circuit at a pressure the same as that used for the actuators without undue power consumption.

The Dreloba system uses a double-diaphragm digital element illustrated in fig. 1.16. The areas on which the control signals are applied to the outer sides of the piston are twice as large as those on the inner sides, so that an extremely versatile and powerful logical element is obtained. The switching time is about 1 ms, and a life of 10^9 cycles is claimed. It is used in conjunction with a simple rubber-disc *or* gate, also shown in fig. 1.16, and standard blocks are produced in which the amplifiers are combined to perform the various logic switching functions. As many as nine double-diaphragm elements or thirty *or* units may be connected passively in series, so circuits may be extended without excessive air consumption.

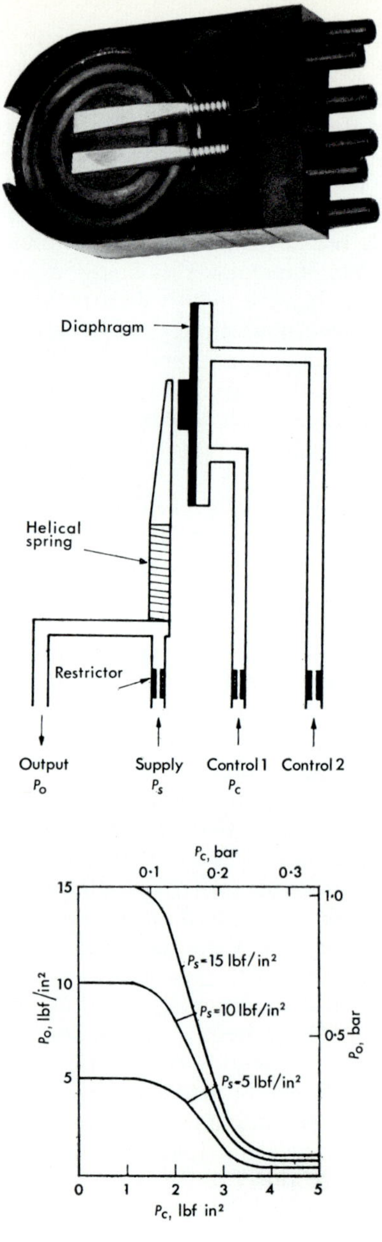

Figure 1.15 Techne *nor* spring unit in fact and principle. Control characteristic shown below.

In a comparison of the various elements the continuous power which is necessary for those with no moving parts may be considered a disadvantage. However, although static elements need power only intermittently, many of them do not take advantage of this feature. Moreover, in practice, leakage through clearances or exhaust apertures may account for a considerable overall power consumption. No moving-parts elements would appear to give freedom from

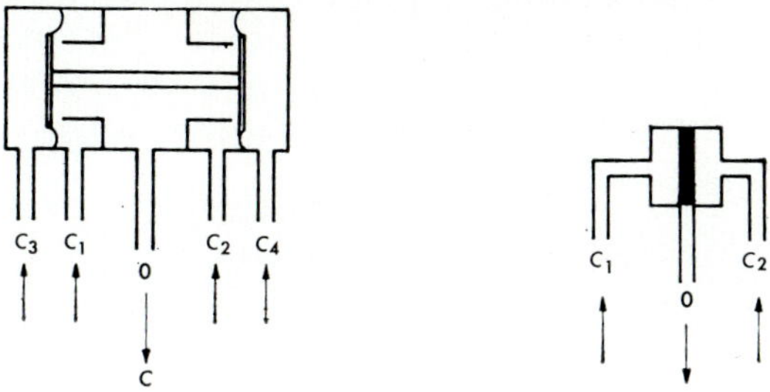

Figure 1.16 Dreloba double-diaphragm and *or* elements. Left—double-diaphragm element. Right—disc *or* element.

C_1	C_2	C_3	C_4	0
0	0	0	0	0
0	0	0	1	0
0	0	1	0	0
0	0	1	1	0
0	1	0	0	1
0	1	0	1	0
0	1	1	0	1
0	1	1	1	1
1	0	0	0	1
1	0	0	1	1
1	0	1	0	0
1	0	1	1	1
1	1	0	0	1
1	1	0	1	1
1	1	1	0	1
1	1	1	1	1

maintenance. However, this is not the same as superior reliability. An element with a guaranteed lifetime, with facilities for regular maintenance and replacement of worn parts, may be preferable to one with a very long but unpredictable lifetime. The reliability of any type of system depends on its susceptibility to various random internal and external disturbances and must be determined by test.

1.10 The influence of fluidics

In this first chapter we have tried to give a guide to some of the important fluid mechanics which underlies the operation of fluid amplifiers. The examples have been chosen from a wide range of element types to illustrate the application of these principles, as well as to introduce some of the important terms of the art. Further types of element and problems of application are described in later chapters. It is clear from the range of these applications that fluidics will develop into some kind of established technology within the subject of pneumatic and hydraulic control. Whether it will be as large and important as some of its advocates suggest is still a matter for conjecture, but one thing is certain: activity in the whole field of fluid control has been stimulated to a degree which gives the industry a new look, whether it continues with or without fluidics.

2

DEVICES AND THEIR MANUFACTURE

Dr A. McCabe
Stockport College of Technology

R. E. Hazard
British Aircraft Corporation (Operating) Ltd

Active fluidic devices are either *digital* or *proportional.*

The flow through *digital* elements with two output receivers is such that it selects one receiver to the exclusion of the other. Digital devices with a single receiver have flow to the receiver diverted suitably, or the state of the flow altered to lose output signal. In each instance the devices have distinct output states and are said to have *on/off* switch characteristics. The range of wall-attachment amplifiers, the turbulence amplifier and the focused-jet amplifier are examples of digital devices. The wall-attachment amplifier and the focused-jet amplifier employ the ability of a jet to attach itself to an adjacent wall, or the Coanda effect, as their basic operating principle. The turbulence amplifier uses an input signal to alter the nature of the power stream and consequently reduces appreciably the strength of the output signal.

The output of a *proportional* or *analogue* device can assume an infinite number of positions in response to small changes in input signal. The output effectively amplifies the pressure and flow signals at the inputs. The beam-deflexion amplifier and the impact modulator use control pressure and flow to deflect a fluid jet. The vortex amplifier utilizes the pressure change across a vortex-sink flow to modulate the output signal.

As well as depending on fluid dynamic phenomena, all fluidic devices are influenced by their geometric configuration. The geometry may be planar or axisymmetric. Most devices have planar form. The advantages this offers in manufacture are described fully below, but it was also adopted as a means of simplifying design because the fluid flow could be treated as wholly two-dimensional. Later work has shown important three-dimensional phenomena, and hopes of obtaining meaningful data from solutions of the general Navier–Stokes equations are small. Device design has mainly been developed by cut-and-try methods. The wall-attachment amplifier and the beam-deflexion amplifier have planar geometries. On the other hand, the focused-jet amplifier, the impact modulator and the original design of turbulence amplifier are examples of axisymmetric devices.

The fluidic devices described below are those which appear to be the most important in terms, not only of present applications and availability, but also of potential for the future. Wall-attachment and turbulence amplifiers have been used in a wide range of applications. Most other fluidic devices have not been generally applied, perhaps because they are proprietary designs or because detailed performance information is lacking. The fluid dynamics of the various devices is dealt with in Chapter 1.

2.1 The turbulence amplifier

The fluid dynamics of the turbulence amplifier (*1*) is discussed in Section 1.6. The amplifier operates at a supply pressure, typically 1·5 kN/m^2, which is low compared with that of most other fluidic devices. A laminar stream is projected across a vented chamber towards a receiver on the power nozzle axis (see fig. 1.7). Only a small amount of the power stream energy is dissipated by viscous effects, and pressure recovery at the output is high. Small disturbances of the power stream are induced near its point of issue by control flows through one or more signal ports normal to the power stream axis. The disturbances are amplified by viscous effects, and the power stream becomes turbulent at a point upstream of the receiver. The result is a significant decrease in output signal magnitude.

The amplifier functions as a digital *nor* element. As the *nor* gate is a basic logic element, devices for other functions can be constructed by interconnecting turbulence amplifiers. Examples are bistable or flip-flop gates for memory functions, *or*, *and* and *nand* gates, and binary counters.

The inherent isolation between inputs and output has made the interconnexion of turbulence amplifiers a relatively simple matter. This facility, plus low cost, has led to the use of turbulence amplifiers in many control applications.

For most of these applications amplifiers of an early form have been assembled. This early version is a soldered tubular structure with a fan-in ratio of four, and pipework must be used to make circuit interconnexions. Details of the production specification are given later in the chapter, in the discussion on device manufacture. However, tubular construction is being superseded by a planar configuration with a fan-in of two, as described by Verhelst (*2*), which is more adaptable to integrated circuit or modular circuit assembly. The fan-in ratio of the planar device is not restricted to two: indeed, a fan-in of six can be achieved relatively easily. The fan-out ratio of both types of turbulence amplifier is typically six.

The switching time of the turbular turbulence amplifier is about 7–10 ms, but times of 1–2 ms are claimed for the planar device.

2.2 The wall-attachment amplifier

The wall-attachment amplifier (cf. Section 1.7) was developed initially at the Harry Diamond Laboratories. It utilizes the phenomenon of jet attachment—a planar jet issuing from a nozzle will attach itself to an adjacent wall. A free jet from a nozzle or orifice entrains fluid from the surroundings. A balance is reached such that the mass flow increases in the downstream direction, the jet width

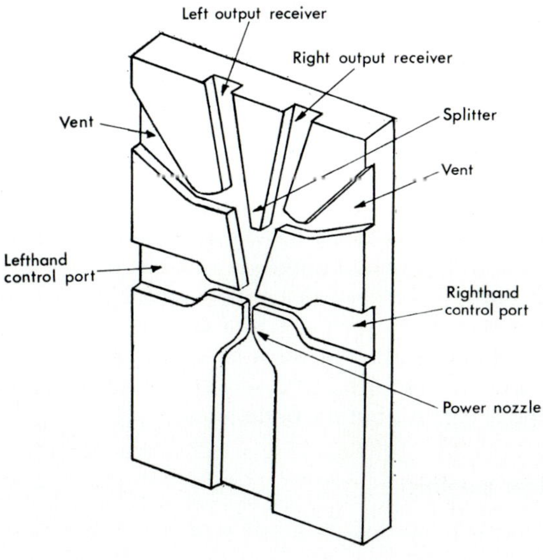

Figure 2.1 Typical wall-attachment element.

increases, its velocity decreases and the total momentum remains ostensibly constant.

When a wall is placed adjacent to the jet, fluid between the jet and the wall is entrained more rapidly than it can be replaced by counterflow from downstream. The removal of fluid by entrainment causes the pressure to fall, the pressure gradient lying transverse to the jet flow. Because of the pressure gradient the stream is deflected towards the wall, thus impeding the counterflow from downstream. The pressure between jet and wall decreases further, and the cycle of events continues until stability is reached when the jet attaches itself to and flows along the wall, as illustrated in fig. 1.3(b). Once the jet has attached itself to the wall, the transverse pressure gradient holds it in position as long as the stream flows.

A typical wall-attachment bistable amplifier profile is shown in fig. 2.1. The jet issuing from the power nozzle attaches itself to one of the adjacent walls by the mechanism described above. The splitter and vent maintain flow stability. If the flow has become attached to the right-hand wall, say, the whole of the available flow and pressure are captured by the right-hand receiver. Flow through the control ports on either side of the power nozzle provides the means for switching the power stream from one output to the other. To switch the flow to the left-hand output, fluid is introduced into the right-hand control port at a rate and pressure which are low compared with those of the power stream. The control input increases the entrainment rate in the low-pressure 'bubble', and the power jet detaches itself and switches to the left-hand wall. When the control input is removed the power jet remains attached to the right-hand wall until a control signal is applied at the right control port.

The ability of the power stream to maintain flow to a particular receiver is termed *memory* or *bistability*. Of course, it persists only so so long as there is a power stream available. The principle of wall attachment is not restricted to the bistable amplifier. Devices have been designed to perform many functions, such as inversion, *or-nor*, *nor*, *and-nand* and binary counting.

The fan-in ratio of memory elements is generally restricted to one or two, but *or-nor* gates with a fan-in of four are available commercially. The fan-out ratio of most devices is about four, and the switch-time, i.e. time required to divert the power stream from one receiver to the other, is about a millisecond.

2.3 Focused-jet amplifier

The axisymmetric focused-jet amplifier (*3*) is shown schematically in fig. 2.2.

The power jet is formed by an inwardly directed annular nozzle of high aspect ratio, where aspect ratio is defined as the ratio of nozzle perimeter to annular exit width. The jet adheres to the flow separator surface, forming a so-called *focused jet*. Fluid flows towards the centre of the flow separator where it is deflected in the axial direction towards a single central receiver placed very close to the flow separator.

Control flows are admitted into a central control chamber and then accelerated by a second annular nozzle surrounding the flow separator. Control inputs detach the power stream from the flow separator. It is diverted from the receiver towards the vented region, and the output signal at the receiver is reduced substantially.

The device is a *nor* element giving a high output when the control is zero and a low output when a control signal is applied. It has a

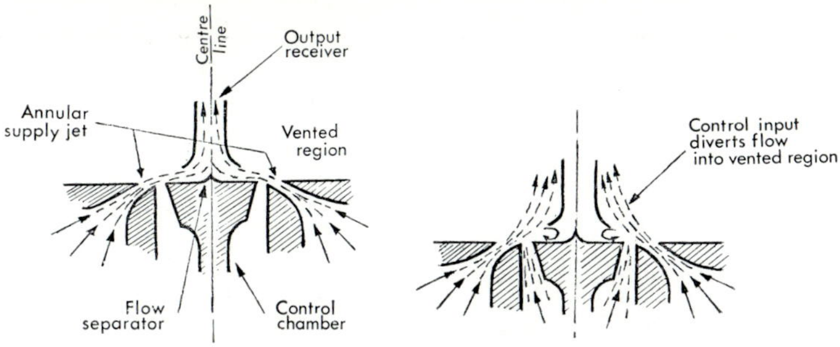

Figure 2.2 Focused-jet amplifier.

fan-in ratio of three or four and a fan-out of six (*4*). The device has good input-output isolation. The interconnexion of devices is a simple matter.

2.4 Proportional devices

2.4.1 *Beam-deflexion amplifier*

The beam deflexion amplifier (cf. Section 1.4) is another device developed at the Harry Diamond Laboratories.

A typical amplifier profile is shown in fig. 2.3. The walls of the interaction region are set back to preclude wall-attachment and to allow adequate venting. In the absence of control flows the high-energy power-jet flows across the interaction region, strikes the splitter and divides evenly between the two receivers. If equal impedances are placed downstream of the output ports the differential pressure between the outputs is zero.

Control or low-energy fluid is directed into the interaction region by nozzles on either side of the power jet. A pressure difference, and hence a flow difference, between the control ports deflects the power jet. It is deflected from the centre-line in the direction of the weaker control signal, and a greater proportion of the resultant jet enters the receiver on that side. Consequently, a non-zero pressure-difference is registered between the outputs. A small power-jet deflexion is sufficient to change the differential output signal significantly.

The degree of deflexion of the power jet is determined by its momentum flux and the forces exerted on it by the control streams. The control jets apply pressure forces and momentum-flux forces: it is usually difficult to say which type of force predominates.

As well as stabilizing the power jet, the vents minimize the effects of output loading. These are the effects of nozzles and/or orifices, connected downstream of the amplifier, on flow entering the receivers. The vented regions are held at a constant pressure, normally the atmospheric ambient pressure, and fluid which cannot be accommodated by the receivers spills over and exhausts through the vents.

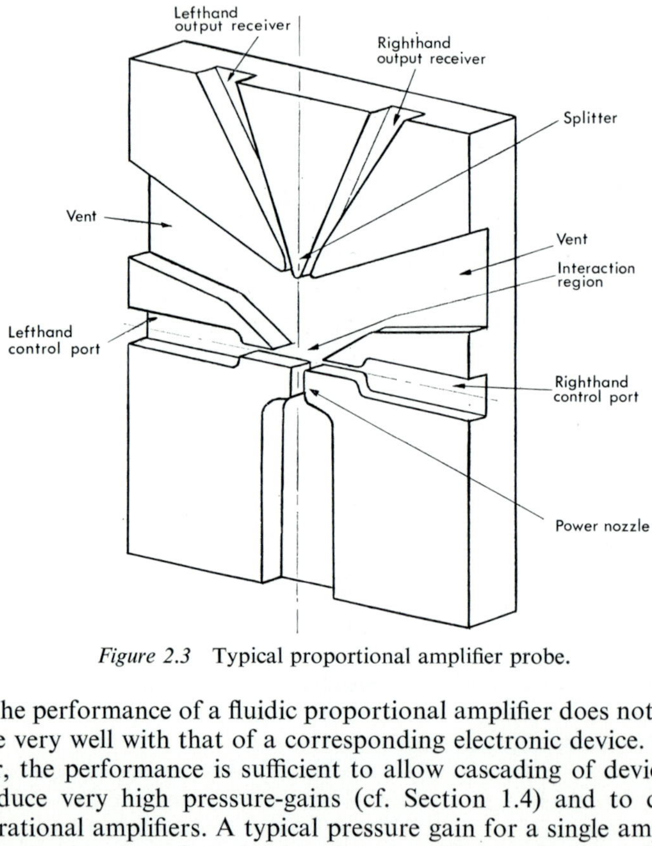

Figure 2.3 Typical proportional amplifier probe.

The performance of a fluidic proportional amplifier does not compare very well with that of a corresponding electronic device. However, the performance is sufficient to allow cascading of devices to produce very high pressure-gains (cf. Section 1.4) and to design operational amplifiers. A typical pressure gain for a single amplifier operating into an infinite load is about ten. The bandwidth, which is a function of amplifier geometry and loading, is normally less than 1000 Hz. The maximum pressure recovery is about 70% of the power-jet supply-pressure and the signal/noise ratio is low.

2.4.2 *The vortex amplifier*

Several variations of the basic vortex amplifier have been suggested and tested, but the amplification mechanism is common to all the devices. It is described in Section 1.8. (See also Sections 5.2, 6.2 and 7.4.)

With no control input the power stream flows directly towards the central receiver (see fig. 1.14). It suffers little impedance other than that encountered by the flow in negotiating a 90° bend. When a control flow is introduced, the power stream takes up a spiral path and the flow pattern in the chamber resembles a vortex-sink flow. As the control pressure is increased, the tangential control flow applies a greater swirl to the power stream, and the swirling flow produces a large pressure drop across the amplifier. The pressure drop is influenced by two factors, first the increase in flow path length and secondly the increase in energy dissipation when the fluid follows a curved path. The magnitude of the pressure drop is proportional to the applied control pressure. With increasing control pressure the total flow through the device is reduced: a minimum is reached when the power stream is shut off completely, and then all the flow passing through the amplifier is supplied from the control port.

The major advantages of the vortex amplifier are a simple geometry, high pressure-recovery, and high fan-in ratio achieved by spacing control ports around the curved wall. Its major disadvantage is that the control pressure must be higher than the supply pressure. The control pressure required to initiate throttling depends on the control- and power-stream flow-rates and the geometry of the device, but it is usually between 25 and 75% greater than the supply pressure.

2.4.3 *Impact modulator*

The direct-impact modulator is an axisymmetric device employing two collinear opposed jets issuing into a vented region (fig. 2.4). An impact plane normal to the centre-line of the power jets is formed. Initially the power-jet pressures are adjusted so that the impact plane

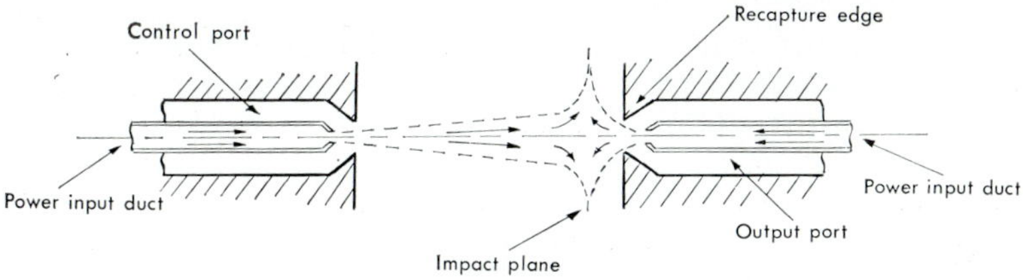

Figure 2.4 Direct impact modulator.

lies very close to the recapture edge of the annular receiver. A control input increases the intensity of the left-hand power-jet, the original jet balance is disturbed, and the impact plane moves to the right. The recapture edge intercepts a portion of the impacted flow and the output pressure rises. Subsequent increases in the control signal increase the output signal proportionally.

The impact modulator provides the highest pressure-gain available from any existing device. Pressure gains as high as 200 (*5*) have been reported, but information on the dynamic performance and signal/noise ratio have not been made available.

In fig. 2.5 a similar device known as the transverse impact modulator is sketched. In this instance the control nozzle is mounted close to the left-hand power nozzle and perpendicular to the axis of the impacting jets. In the absence of control signal the impact plane is sufficiently close to the recapture edge for some of the outwardly

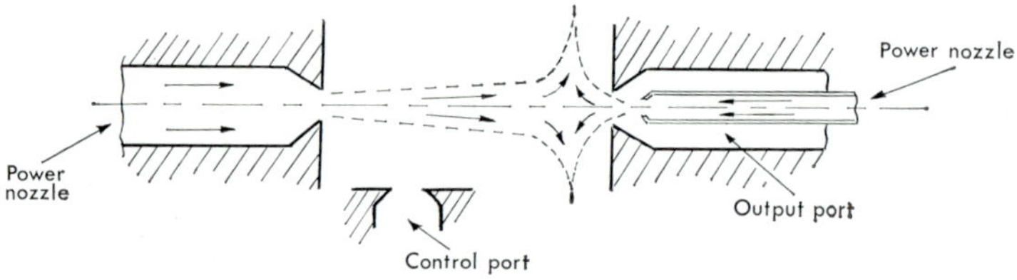

Figure 2.5 Transverse impact modulator.

directed flow to be collected by the output receiver. This portion of the flow is known as the *t.i.m. reference output level* (*6*).

Application of a control input reduces the intensity of the left-hand power-jet by deflecting it away from the power-jet axis. Consequently the impact plane moves to the left and the amount of fluid captured by the output receiver is reduced. The output pressure varies inversely with the control input pressure, and the device has a negative pressure gain.

Again, only limited information is available on the performance, but no-load pressure gains of -100 have been reported.

2.5 Design requirements

The foregoing paragraphs have outlined various fluidic devices that have been analysed by research and development workers. In this research the working fluid has mainly been air because it is commonly used in many sections of industry. Therefore the specification requirement has been written to suit this, the largest potential market, having in mind standard amplifiers less than 40 mm square in plan, fluid pressures below 100 kN/m^2, and jet speeds which are subsonic.

SPECIFICATION

Shape accuracy (when used for control circuits)	Generally speaking the critical dimensions must be held in a tolerance band approximately $\pm 3\%$ about the nominal, to cater for a further change in shape with age.
Quantity reproduction accuracy	Manufacture should not rely on craft skills (except in making tools, jigs and

fixtures) and the main production should be by automated machinery.

Surface finish — A smooth finish is desirable in an amplifier, but is not so critical as dimensional accuracy. Both are directly related in manufacturing processes and one finds that the control of accuracy suffices to control finish.

Dimensional stability — Having ensured high accuracy it is necessary to choose stable materials that will withstand environmental effects during production and in use. A good amplifier should withstand (a) handling, (b) damage by the fluid, whether by erosion, chemical attack, absorption or clogging, and (c) temperature variation.

Leakage — A rough guide: interchannel or channel-to-atmosphere leakage should be less than 1% of the supply flow.

Cost — In order to allow circuit designers a maximum choice of control or logic elements, any technical advantages of fluidic designs must not be lost by a non-competitive price.

Working life — A life expectancy of not less than 10 years, and preferably of 25 years, is desirable.

2.6 Manufacture

The first pneumatic fluidic device to be produced for the industrial market by a British manufacturer was the turbulence amplifier, a design utilizing a laminar jet (see fig. 1.7). The design was initially made, and not surprisingly, in a traditional manner by jigging brass tubes carefully into position and soldering them together. Presumably the aim was to meet the specification by:

- using accurately sized tube to obtain a laminar supply jet, correct receiver size, and correct signal-port size
- using small-bore smooth tubes to obtain consistent fluid flow
- accurately positioning the supply jet relative to the receiver to ensure concentricity and alignment during and after soldering
- soldering to leakproof the unit
- accurately positioning tubes to obtain repeatable amplifier performance
- using materials which give stability of shape and resistance to erosion, corrosion and absorption
- using materials which enable the amplifier to work over a large temperature range without significant change in shape
- working to a realistic cost related to a viable sales campaign

In the meantime the wall-attachment amplifier (see fig. 2.1) and the beam-deflexion amplifier (see fig. 2.3) were attracting a lot of attention because they offered, particularly to the aerospace industry, more attractive performance. But their designers were asking for rectangular nozzles with aspect ratios of two or more. This requirement, in the context of the needs of research and manufacture, led to the evolution of the *planar* shape. The contributory factors were:

- The advantage of a black-and-white drawing as the first-stage tool in manufacture, and the use, as the production reference, of glass negatives reproduced from the black-and-white drawing.
- The utility of Dycril for research models.
- The technique of copying a Dycril model and producing epoxy cast amplifiers for small batch production without need for *costly* tooling.
- The possibility of injection- or compression-moulding the two parts of an amplifier when costly tooling for mass production is warranted.
- The fact that amplifiers can be built from thin metal foils for high-temperature models. Here the manufacturing technique is again photographic copying and chemical etching.
- The manufacture of glass ceramic amplifiers, also with the aid of copying and etching techniques.
- The design of amplifiers, breadboard builds for prototypes and integrated circuits for production.

2.6.1. *Black-and-whites*

The traditional vehicle for taking information to a workshop is a detailed drawing of the shape required, fully dimensioned, toleranced, and quoting material and finishes. The workshop considers the manufacturing techniques necessary to produce the components in the quantities required.

From such investigations as those described in the preceding section there have been developed manufacturing techniques which necessitate both the traditional detailed drawing and a first-stage tool which is also a drawing. Close liaison is obligatory between the design and production sections, and usually the design office prepares the tool drawing.

To attain the desired accuracy it is best to use a Melinex-based clear draughting paper, which has good stability (fig. 2.6). On this one can stick black tape to delineate the plan form of the fluidic amplifier at least twenty times full size. The shape will not be a true enlargement of the amplifier because allowances have to be made for the effects of the manufacturing processes which are to follow, and it is prudent to state this knowledge on the black-and-white.

The drawing office black-and-white is photographically copied and reduced to suit the next phase of manufacture. When a production reference is required the drawing office picture is reproduced to the correct scale on a glass plate, with reverse form. If not, a negative is made on Cronor D photographic film, a grade of thin film which gives an excellent copy with high contrast (see fig. 2.7).

Negatives can now be made from either of these references, again on Cronor D photographic material, to give the correct masking

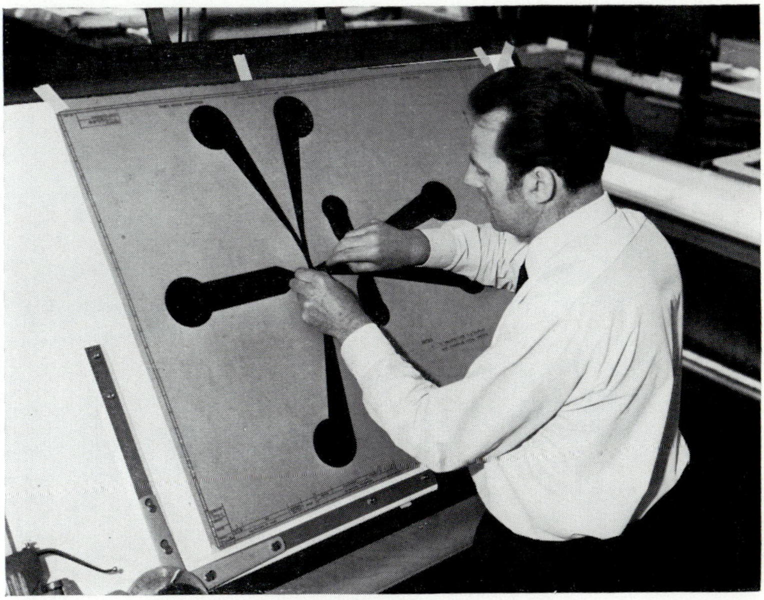

Figure 2.6 Draughting a black-and-white.

shape. This photographic work is done in two stages only to simplify the work on the twenty-times-full-size black-and-white.

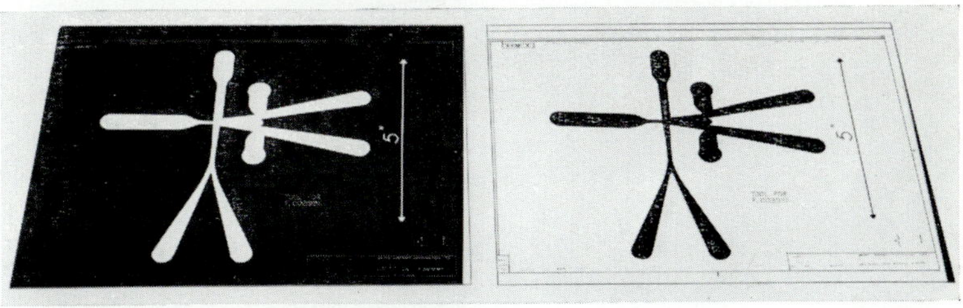

Figure 2.7 Left—Cronor D film. Right—glass plate.

The film finally produced is the workshop tool for the next phase of production. Its life is limited, and glass plates must therefore be kept from which further copies can be taken if full-scale production is contemplated.

The next phase, irrespective of manufacturing technique, is based on photographic copying and the use of light-sensitive materials. Let us explore the most important of these, a technique which made possible the rapid development of amplifier designs.

2.6.2 *Dycril amplifiers for research*

Before research on fluid amplifiers had started in Britain the printing trade were using a material called Dycril from which they could

Figure 2.8 Dycril under exposure to light.

produce printing plates. This material, a product of E.I. du Pont de Nemours and Co., consists of a thin layer of transparent, light-sensitive plastic, bonded to either a rigid or a flexible base.

It was found that one type of Dycril, known as 152 Rigid Flat, had a 1-mm depth of photopolymer plastic and that this could be etched to give channels down to $\frac{1}{2}$ mm wide, or, when very special care was taken, down to $\frac{1}{4}$ mm wide. This size range was very fortunately compatible with designers' requirements, and the use of Dycril for research models allowed the development of planar amplifiers, both digital and analogue, to proceed at a very rapid rate.

To make a Dycril amplifier one uses unexposed material and the appropriate thin photographic film. The photographic negative produced by du Pont is preferable because that company has specially developed it for masking Dycril. The matt side of the film is placed in contact with the clean piece of Dycril, and together the pieces of film and Dycril are inserted into an ultra-violet exposure frame. The space between the pieces in the frame is evacuated to ensure that film and Dycril are in intimate contact before exposure (fig. 2.8)

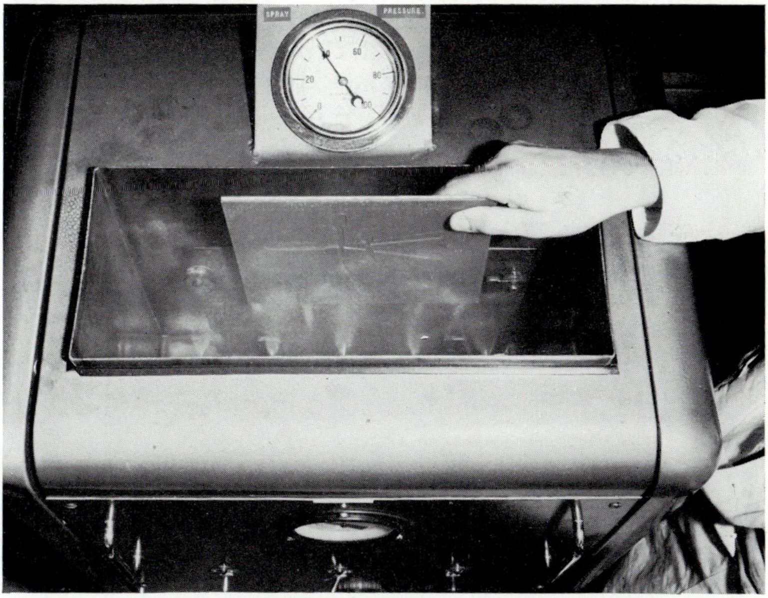

Figure 2.9 Dycril being washed.

to ultra-violet light. The u.v. light is suitably collimated, and incident at right-angles to the Dycril and negative. Material masked from u.v. light by the negative stays in the soft, monomeric state, while exposed material hardens into the polymerized state. Exposure times are not more than 20 min long, depending on the equipment.

The Dycril is hidden from light and placed as soon as possible in a dark spray booth, where the soft plastic is dissolved away (fig. 2.9).

For this operation a solution of $\frac{1}{2}\%$ sodium hydroxide in distilled water is heated to 40°C and pumped through nozzles as a strong spray. The caustic solution softens the plastic and the jets drive the soft plastic out of the channels.

Because Dycril absorbs water great care must be taken when producing the accurate channels. Sometimes it is necessary to dry out the Dycril between washes to obtain the desired results.

When the Dycril is dry the result is half an amplifier. There are many methods for fitting a cover and connecting pipes to it before testing the device in the laboratory. We prefer to lap the top surface of the Dycril and fix onto it, with double-sided sticky tape, a Perspex cover into which go the necessary vents and connectors.

2.6.2 *Epoxy amplifiers for small-batch production*

Amplifiers made from Dycril are only suitable for research. They must be handled with great care and kept in a dry, warm atmosphere to keep their accuracy. They are not practical for industrial applications because the plastic is very soft, changes shape dramatically in a humid atmosphere, and is easily attacked by oils and greases.

However, it has been found possible to copy the Dycril model by means of the following method. It was developed from our experiments with cast materials during the evolution of potted electronic

Figure 2.10 Mould for silicone rubber.

equipment. The inspiration and the clue to success was the use of silicone rubber as the first-stage mould. The rubber overcomes the release problem raised by the undercutting of the Dycril channels, i.e. the channels are wider at the bottom than the top.

A mould (fig. 2.10) is made from an acrylic plate into which a sealing edge is cut and to which screws and metal inserts are fixed. The inserts are treated to ensure a good key with the silicone upon curing. The surrounds are made of plastic strips and stuck together and to the Perspex base with beeswax. A pre-machined Dycril or metal master (fig. 2.11) of correct profile, and with a port arrange-

Figure 2.11 Metal master.

ment which meets the performance requirements, is stuck to the Perspex plate inside the box with double-sided sticky tape. To obtain a second-stage silicone-rubber mould, silicone rubber made from two thoroughly mixed liquids is poured into the finished mould. The assembly is then put under vacuum to remove air bubbles and to draw the fluid into the amplifier channels.

Once the rubber has set, the fixing screws and box surrounds can be removed and the rubber peeled carefully away from the Dycril. Note that the tenon joint can only be broken because the rubber is elastic. Pouring and riser vents can now be cut in the rubber. The Perspex top (minus the Dycril amplifier) is added, and the cavity is ready to be filled with epoxy resin (see fig. 2.12).

To produce bubble-free castings in epoxy resin takes a little skill. It is found best to mix the liquids, warm them to reduce their viscosity, remove as much air as possible under vacuum, and then pour the liquid carefully into the moulds. When full the moulds are left for the resin to cure. It is interesting to note that when silicone moulds are used no release agent is necessary.

Figure 2.12 Silicone mould and plastic lid.

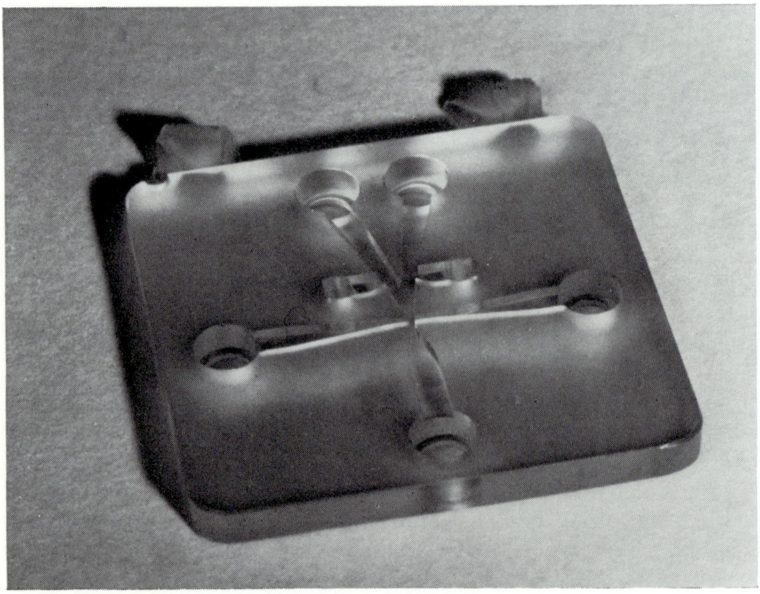

Figure 2.13 Half-element.

When the amplifier half (fig. 2.13) is removed from the mould it is trimmed and the mating face prepared to receive a top cover. This poses another difficult problem, the choice between:

Cementing the halves together
Riveting the halves together, with or without a gasket
Moulding on the cover using the lost-wax technique.

Decision is difficult because each method has disadvantages. The lost-wax process was the most successful in early development, and it produced an elegant amplifier.

The amplifier channels are filled with an 'investment casting' wax (fig. 2.14). The top face is carefully cleaned so that no wax is

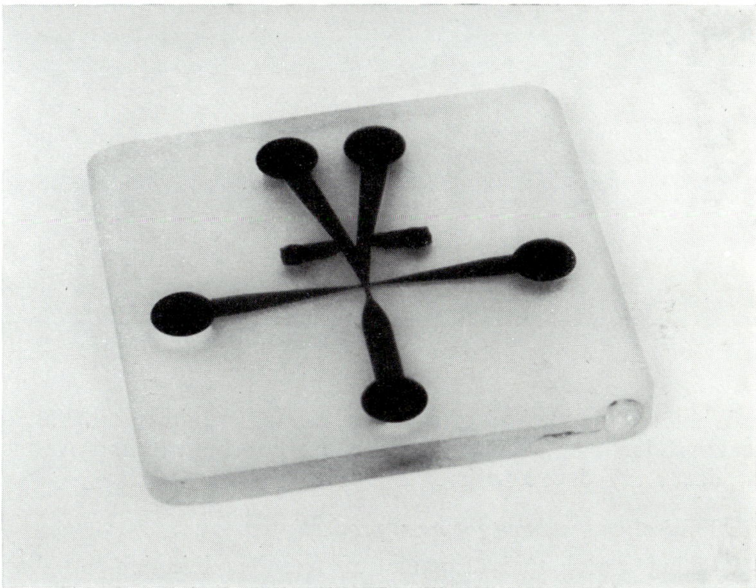

Figure 2.14 Element with wax.

left on the top epoxy face, only in the channels. This half is fitted into another silicone mould and the second casting phase is completed in exactly the same way as the first. The wax is then removed by heating, preferably in a trichlorethylene degreaser. Any further treatment is simply for saleability (fig. 2.15).

Stability can be improved by including a filler in the epoxy resin, but clear epoxy has the advantage that one can see the channels. This is useful during testing or fault-finding.

Covers can also be cemented into position but it is difficult to stop cement finding its way into a channel. One method to obviate this is to use a thin paper carrier. The paper is impregnated with epoxy resin (the cement) and laid onto the cover. Then the half-amplifier is put into position. Light pressure will now ensure a good bond.

Figure 2.15 Epoxy cast element.

Whatever the method of covering, casting requires skill. Large-scale production warrants a de-skilled method, both to improve the product and reduce its price.

2.6.3 *Injection moulding for mass production*

Making metal tools capable of resisting injection-moulding pressures and temperatures is very difficult. Their employment calls for careful design and good use of previous experiences. For instance, it is an advantage to:

- keep the aspect ratio of the amplifier to a minimum
- use a Dycril master from which to copy-mill
- make a female form by milling where possible
- use the nickel plating process.

Copy-milling on an engraving machine, following a Dyçril master, has two advantages. It makes use of the known technique for producing an accurate Dycril, and sharp corners are made in the female form where required by the milling process.

If the aspect ratio is small it is possible to nickel-plate a plastic female form which has been copied on an engraving machine.

If the aspect ratio of the amplifier is such that nickel plating is impossible, the male form may still be copy-milled, for it can be trimmed manually.

Obviously the amplifier cover can be made as a moulding quite easily, but the manner of joining calls for investigation.

2.6.4 *Metal amplifiers for high temperatures*

The metal etching technique for thin foils, already fully developed for such items as printed circuits, can also be exploited successfully.

This process also employs a black-and-white photographic negative mask as previously described for use with Dycril. The difference here is that the metal foil is covered by a photo-sensitive material, known

Figure 2.16 Metal foil amplifier.

as a *resist*, which becomes stable after exposure to ultra-violet light. Areas masked during exposure are left soluble and can be etched away. The etching chemical attacks both the metal and the soft resist.

Unfortunately etched channels do not have vertical sides. To obtain the desired results it is best to use thin foil and attack it from both sides. Attacking the foil from both sides has the advantage of halving the variation of channel width, but the advantage is only gained up to the limit of mismatching.

Figure 2.16 shows a typical metal foil amplifier.

2.6.5 *Glass ceramic amplifiers*

Corning Glass of the U.S.A. have developed a glass which is chemically changed by exposure to ultra-violet light. After heat treatment there is a 25 to 1 activity ratio between exposed and unexposed glass. The exposed material can be etched and the resulting deep channels have been found to have the required accuracy. Thus the technique has similarities to the etching of Dycril and resist-covered metal foil.

Unfortunately it is not possible to buy the material on the general market and its production is covered by patents.

2.6.6 *Large amplifiers*

When amplifier nozzle sizes are larger than 1×4 mm it is unnecessary to use special production techniques. An experienced engineer would find no trouble in determining the most appropriate method.

3

MEDICAL APPLICATIONS

Dr G. A. Parker
University of Surrey

In the last twenty years there has been a tremendous increase in the development of medical technology. This has brought about the development of a wide range of equipment, much of it very sophisticated, to meet the challenges posed by the new advances in medical knowledge. Much of the equipment is expensive and requires highly skilled personnel to operate it satisfactorily. Thus in the face of rapidly escalating costs in the medical field there has been a trend towards automated equipment and cheaper alternative solutions to problems.

How does fluidics apply to medical technology? Fluidic devices have potentially good reliability, are easily understood by unskilled personnel and yet are relatively cheap to produce. In addition the control of fluid processes, either in the human being or in such equipment as automatic analysers, is a general requirement for medical purposes and is well suited to fluidic concepts. A further factor which would appear to make pure fluidic devices very attractive for medical applications is that they would be easy to sterilize and autoclave if constructed from compatible materials. The same considerations would probably also apply to simple moving-part devices.

3.1 Artificial heart-pump

The study of the artificial heart was started in America by Kolff, Akutsu and others (*1–4*). Initially the investigators were concerned with obtaining pulsating flow and using the correct design and materials for the pump to prevent blood coagulation and damage to the blood corpuscles (haemolysis). Since then there has been a continuing quest for better designs, and much effort has been devoted to making the output-pressure waveform of the pump correspond more closely with that of the living heart. Various control media for the pump have been considered, and there has been a tendency to use air for this purpose, even in non-fluidic equipment.

In general an artificial heart-pump may be distinguished from the artificial cardiopulmonary apparatus in that it is intended to perform the cardiac function over long periods rather than purely during surgery. It is therefore of paramount importance that blood clotting and haemolysis be reduced to a minimum in such units.

Artificial hearts in current use may be divided into two categories: the first for full substitution and the second for auxiliary circulation. Full substitution of an artificial for a natural heart is intended to be permanent. Auxiliary circulation methods allow for the treatment of cardiac or circulatory insufficiency *in vivo* for periods up to a month. In the latter case the artificial heart is placed serially or in parallel with the natural heart to assist the cardiac function and is removed after the patient's recovery.

Figure 3.1 shows the different types of assisted circulation configuration. The right heart by-pass method consists in the placement

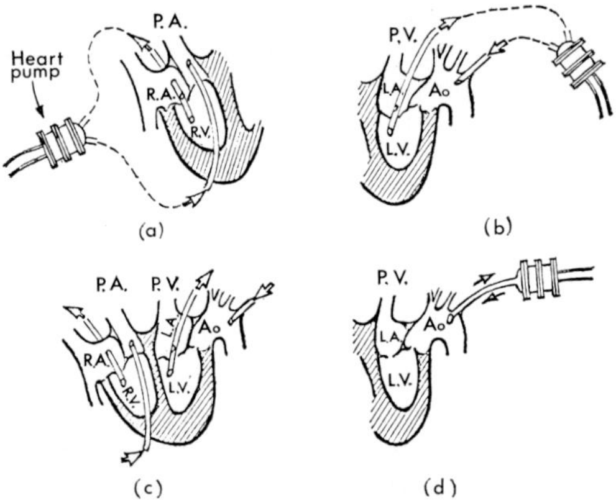

Figure 3.1 Assisting the heart to circulate blood (*8*). The methods shown are: (a) right by-pass, (b) left by-pass, (c) total by-pass and (d) counter-pulsation. Key: A.—aorta, P.A.—pulmonary artery, P.V.—pulmonary venous, R.A.—right auricular, R.V.—right ventricular, L.A.—left auricular, L.V.—left ventricular.

of an artificial heart in parallel with the natural heart *in vivo* to assist the right heart system in supplying blood to the lungs. The left heart by-pass method is used in a similar way to assist blood flow to the rest of the body. However, the counter-pulsation method places the artificial heart serially in the heart artery in order to supply and receive blood from the natural heart in synchronization with the systole and diastole of the heart, thereby reducing cardiac work.

It is interesting that, up to the time of writing, no fluidic systems have been reported which remove the major cause of blood damage, namely the conventional valving associated with blood pumps.

Instead, fluidic systems have provided control of the pumps by using a gas medium, and are claimed to have pumping and haemolytic characteristics at least as good as those of available units while costing considerably less to produce.

An American team has developed a fluidically controlled heart-pump of the collapsible-bag type (5). The basic circuit is shown in fig. 3.2. It consists essentially of a bistable pure fluid amplifier controlling the pressurization of a plastic bag pumping the blood. The basis of the fluidic part of the circuit is a special wall-reattachment element, without normal vent channels so that it is very sensitive to

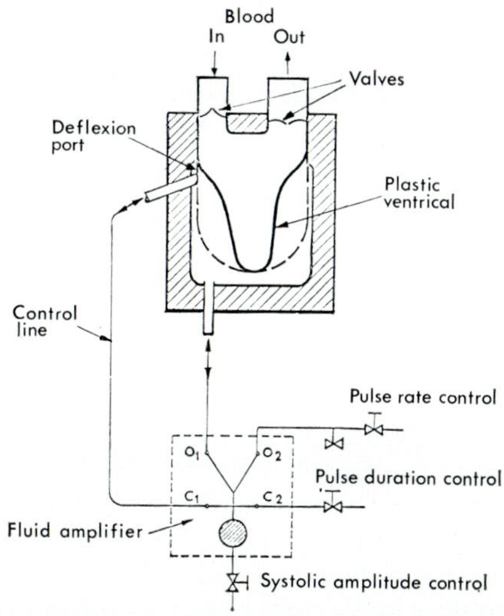

Figure 3.2 Control circuit for heart pump (from *Biomedical Engineering*, Vol. 2, No. 10, Oct. 1967, p. 439).

changes in flow in the output channels. A fluid, usually air, enters the amplifier after passing through the systolic amplitude control valve which regulates the airflow. If the jet attaches itself to the left wall it flows through the left output O_1 into the ventricle housing, where the increased air pressure squeezes the plastic ventricle until the deflexion port is uncovered. The collapse of the ventricle forces blood out through the plastic tricuspid valve, thereby simulating a cardiac systole. With the opening of the deflexion port air flows back into control port C_1 of the amplifier, switching the power jet into the right output O_2.

The ventricle is now no longer pressurized so blood falls into it, thereby causing the deflexion port to be closed. At this stage the supply jet remains switched to output O_2 only as long as the jet can entrain sufficient fluid from the reverse flow down O_1 as the ventricle expands. When this has completely filled, entrainment on the left of the supply jet stops and the pressure across it builds up to a predetermined level preset by the pulse duration control associated with control port C_2. Atmospheric air then flows in and is sufficient to cause the jet to switch back to the left wall, thereby completing the cycle.

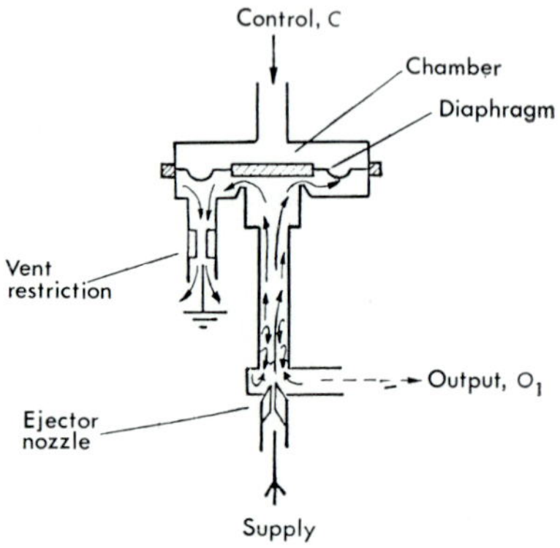

Figure 3.3 Miniature diaphragm switching element (*6*).

The rate at which blood is pumped is controlled by the pulse rate valve open to atmosphere on the right output O_2. This effectively controls the entrainment flow away from the ventricle chamber. Performance data for one experimental pump gives a blood pumping pressure range of 0 to 500 mmHg, corresponding to a flow range of 0·5 to 10 litre/min. The pulse rate could be varied between 30 and 180 count/min.

Brychta (*6*) has used a similar type of control circuit for a heart-pump but claims a substantial reduction in air consumption by the use of moving-part diaphragm devices. The basic features of the element, shown in fig. 3.3, are a diaphragm chamber and an ejector supply nozzle. A cylindrically shaped supply chamber, which acts as an ejector because the output O_1 is taken from a point of maximum negative pressure close to the nozzle exit, is controlled by the position of a diaphragm downstream. When a control signal C is present the diaphragm closes the supply chamber and the output pressure rises to the same value as that of the supply. Removal of the control signal allows the supply flow to vent through a restriction in the diaphragm chamber wall, thereby reducing the output signal to a low value. Suitable choice of geometry determines the output pressure levels and switching gain.

Figure 3.4 shows the circuit used for controlling a heart-pump which has an action similar to that of the American design. The output from the diaphragm control element is high when the diaphragm is closed, which in turn depends on the control line being closed by the plastic ventricle. Pressurization of the bag pumps blood and uncovers the control line. The reduction in pressure in the air chamber surrounding the ventricle is controlled by a suction restriction in conjunction with the pulse-duration control-ejector nozzle. At a predetermined suction pressure the diaphragm snaps open and the ventricle expands owing to gravity flow into the chamber. This eventually causes the control line to be closed by the bag, which increases control pressure sufficiently to close the diaphragm again.

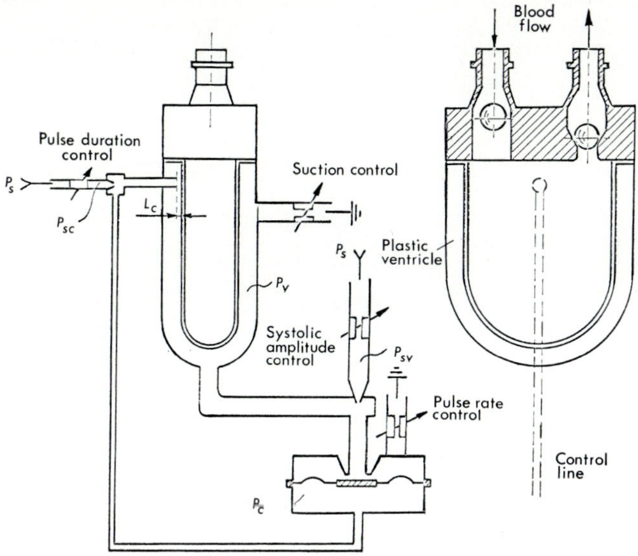

Figure 3.4 Heart pump control-circuit based on diaphragm switches (*6*).

Figure 3.5 shows the system pressures and ventricle displacement waveforms.

Recently Tsuchiya *et al.* (*7, 8*) have developed a Bellophragm-sealed piston heart-pump controlled by a bistable wall-reattachment

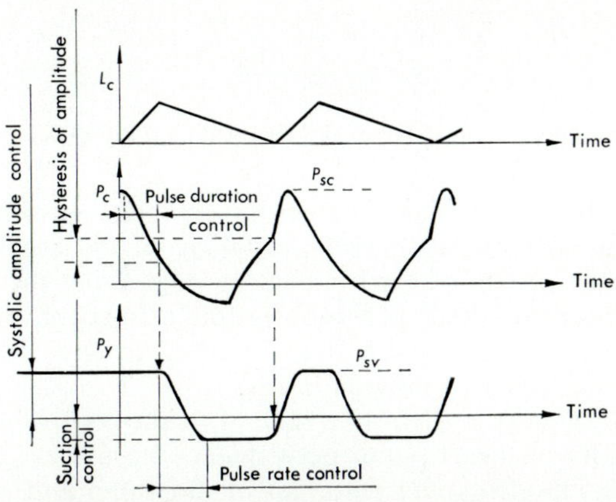

Figure 3.5 Waveforms associated with control circuit shown in fig. 3.4 (*6*).

amplifier. They have conducted animal experiments using the pump for auxiliary circulation. As in the two fluidic heart-pumps previously described, the ventricle chamber forms the feedback volume in an asymmetric oscillator circuit as shown in fig. 3.6. However, there is no triggering of the feedback signal by the pumping piston itself, so the control resistance R_1 largely determines the pump frequency. Manually preset stops adjust the piston stroke to give the desired

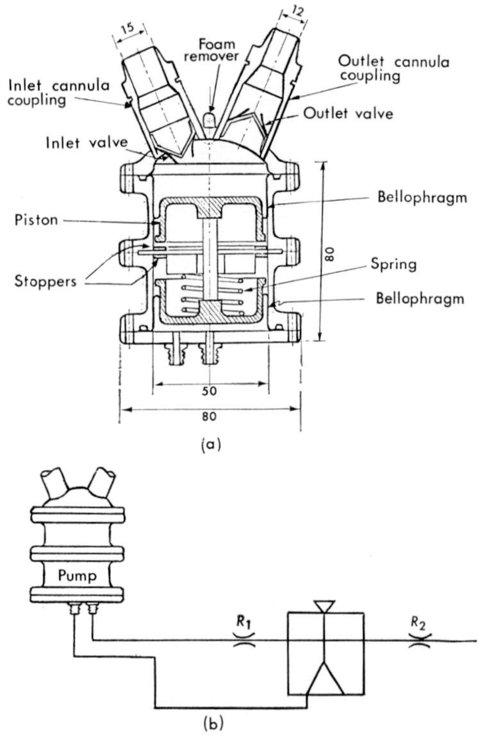

Figure 3.6 Bellophragm blood pump and circuit (*8*). (a) Pump details. (b) Pump circuit.

flow rate, although at high frequencies complete stroking of the pump may be lost owing to early switching of the amplifier. The table shows the results of experiments on dogs using the left heart by-pass method for blood perfusion lasting a relatively long time. Longer-period tests on seven dogs resulted in five deaths. The longest survival time was thirteen hours.

It would seem that in the use to date of fluidic control techniques the main limits on heart-pump performance have been set by the mechanical components and materials of the pump and not by the control system. The reliability and cheapness of the fluidic components used would seem to be more than adequate for this type of application.

LEFT HEART BY-PASS EXPERIMENTS ON DOGS (*8*)

Weight (kg)	*Pumping time* (h)	*Plasma* Hb (mg/dl)	*Result*	*Cause of death*
16	5	—	Died	Ventricle fibrilation
22	6	—	Died	Respiration insufficient
17	11	7 hr 160	Survived	—
25	8	—	Died	Embolism of coronary artery
15	10	7 hr 153	Died	Respiration insufficient
30	11	5 hr 125	Died	Right heart failure
17	13	7 hr 180	Survived	—

3.2 Lung ventilators

Many types of lung ventilator exist. They are often bulky because of their mechanical complexity, which is necessary to provide satisfactory performance for a wide range of patient-lung conditions. Control techniques include the use of electric, electromechanical, pneumatic and magnetic devices which make effective sterilization

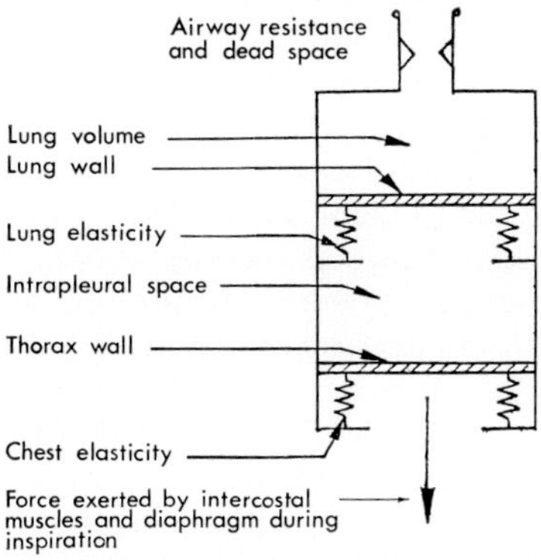

Figure 3.7 Mechanical equivalent of pulmonary system (*12*).

of the ventilator unit difficult. The application of fluidics to ventilators is an attractive idea because relatively simple, cheap, non-moving-part control systems are made possible which would be amenable to autoclaving sterilization techniques.

Assisting or completely controlling the ventilation of a patient by forcing air into and out of his lungs is artificial and only approximates to natural breathing. Figure 3.7 shows a simple model of the human lung system. It contains two volumes, the lung and the intrapleural space. During inspiration the lung muscles cause the lower diaphragm (thorax) to move down, thereby creating a small vacuum in the intrapleural space. This causes the lung wall to move down also, and in so doing induces flow into the lungs from the atmosphere, which is at a higher pressure. The reverse occurs during expiration, when the muscles relax and the thorax and lung wall return to the original positions. Air is expelled through the airway at slightly above atmospheric pressure. Throughout the breathing cycle the pressure variation in the lung is small (a few cmH_2O) since it only has to overcome the airway resistance. The pressure in the intrapleural space is normally below atmospheric throughout the cycle.

Using controlled ventilation the pressure rise in the lung must be sufficient to expand it as the muscles do not perform their normal duty. Typically this work requires a positive pressure of 15–20 cmH_2O during the inspiratory part of the cycle, with a fall to atmospheric pressure during expiration. Interpleural space pressures are significantly increased and may reduce venous blood return, and pulmonary blood flow, and may directly compress the heart. The harmful effect of positive pressurization of the lungs is usually limited by reducing the inspiration phase to a minimum. A generally accepted range of the ratio of expiration to inspiration is 1·5 to 5. Additional design features which improve the ventilator are low airway resistance, small connecting-pipe volumes and the use of negative pressures during exhalation.

There are three basic types of lung ventilator, the *volume-cycled*, the *pressure-cycled* and the *time-cycled*. The volume-cycled type introduces a fixed volume of gas into the patient's lungs irrespective of the pressure or time taken. This automatically ensures that the correct volume of gas, known as the *tidal volume*, is breathed in each cycle. In pressure-cycled systems inspiration and expiration lung pressures are monitored and are used to switch the ventilator on and off at predetermined pressure levels. The time-cycled unit has an inspiratory phase, timed independently of the patient's condition.

Ventilators can be further classified according to whether they operate independently of the patient's inspiratory effort (in which case they are called *controllers*), assist the patient (in which case they are called *assisters*) or combine both functions manually or automatically (in which case they are called *assister-controllers*).

Fluidic ventilators of all three types have been built. There is little evidence that any of them have been subjected to extensive clinical tests. A simple volume-cycled ventilator was first developed in America by Woodward, Joyce and others (*9*, *10*, *11*). It consists of a conventional expanding and collapsing bellows unit controlled by two wall-reattachment bistable fluidic devices. Figure 3.8 shows schematically the circuit developed. Inspiration is brought about by forcing a fresh charge of air into the patient as the bellows are compressed owing to the upward movement of the coupled piston. Pressure is exerted on the piston when the left output O_2 of bistable element A is switched on and the corresponding element B is switched to exhaust O_1. The stroke of the bellows is preset to activate a mechanical trigger on the bottom of the bellows. The trigger switches

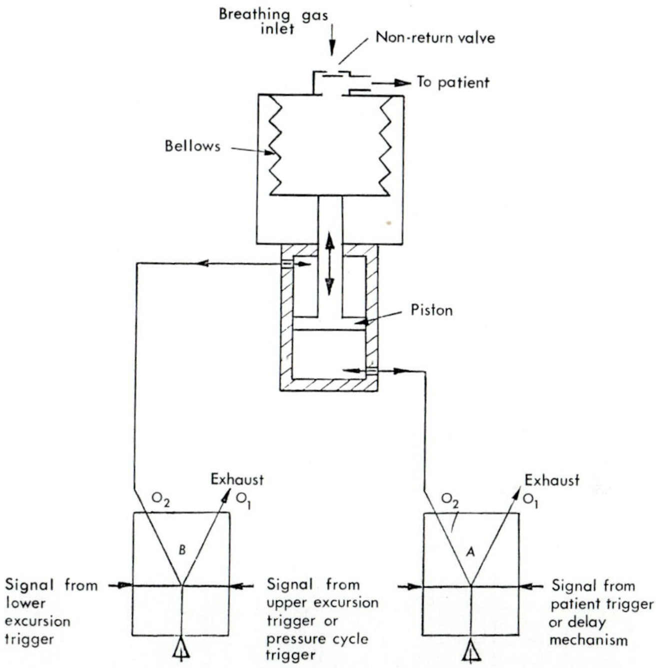

Figure 3.8 Volume-cycled ventilator (*12*).

A to output O_1 (exhaust) and B to output O_2 (pressure), thereby reversing the piston and bellows movement. This allows the patient to exhale to atmosphere through a separate non-return valve (not shown) and also refills the bellows with fresh air through another non-return valve from the supply. At the completion of exhalation a further mechanical trigger is operated. It switches B alone to O_1 and activates a delay circuit. As both amplifiers are now exhausting, the bellows remain in their position until either the preset time delay signal is given (controller mode) or the patient initiates an inspiration pressure switch (assister mode). Data presented by Woodward *et al.* (*9*, *10*, *11*) give a frequency range of 10–60 c/min, maximum inspiration pressures up to 50 cmH_2O, tidal volumes in the range 300 ml–1000 ml, and patient trigger sensitivity in the range $-0{\cdot}5$ to -5 cmH_2O. Air consumption is 3 ft^3/min at 30 lbf/in^2. Although the design significantly reduces the number of moving-part devices, further simplification of the sensing function is possible.

Ball (*12*) has described a more sophisticated fluidic control system

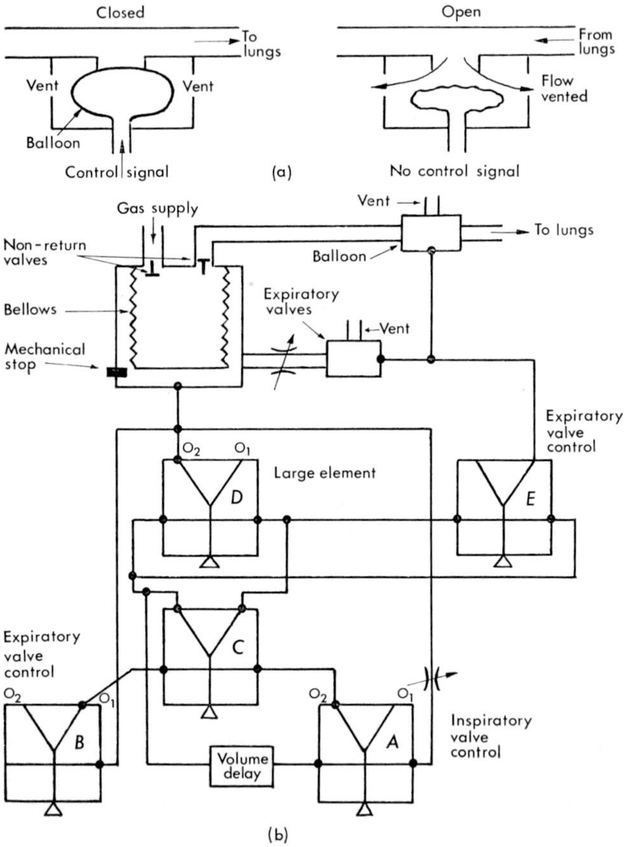

Figure 3.9 Volume-cycled ventilator, after Ball (*12*). (a) Balloon valve details. (b) Circuit details.

for a volume-controlled ventilator in which fluidic pressure switches are used to monitor the bellows chamber pressure to control inspiration and expiration. Figure 3.9 shows the circuit used. It is based on bistable wall-reattachment fluidic elements. During inspiration the bellows are compressed until limited by a mechanical stop preset to control the tidal volume delivered. At this point the chamber pressure rises rapidly until the feedback signal to element *A* is sufficient to switch the output to O_2. This in turn switches the outputs of buffer amplifier C, which vents the large amplifier D through O_1. The patient is now in the expiration phase so the output signal from C is used to trigger a further bistable element E which deflates a balloon valve allowing air from the lungs to escape to the atmosphere. The same signal from C also resets A after passing through a delaying volume. A similar pressure control arrangement is used to switch the start of inspiration after the bellows chamber pressure has dropped to a predetermined level. The suction switches element B from output O_2 to O_1, which causes output element D to switch to output O_2, thereby re-pressurizing the bellows unit. At the same time

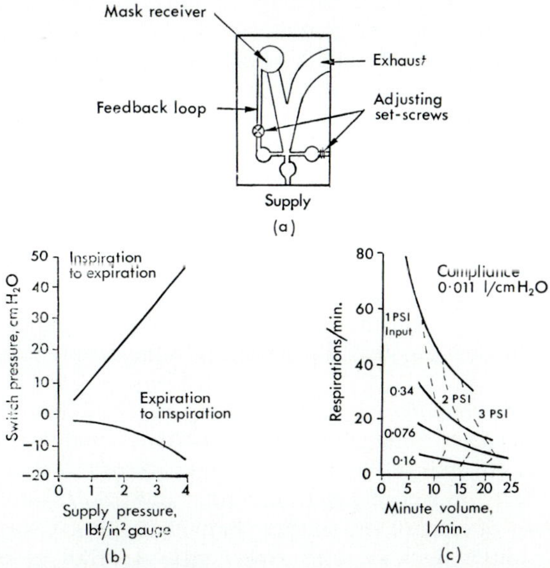

Figure 3.10 Time-cycled ventilator, after Strauls. (a) Ventilator configuration. (b) and (c) Performance characteristics.

element E also changes state and pressurizes the balloon valve to close off the expiratory valve. The time to vent the chamber could be considerably reduced by a similar balloon valve on the bellows chamber, and allow the ratio of expiration to inspiration times to be varied from 1·5 to 4. The advantages claimed for this system are that the tidal volume may be preselected, it is insensitive to small changes in the condition of the patient's lungs, and the inspiration/expiration ratio can be varied. In addition the cycling frequency can be varied by controlling the flow to the large-scale amplifier D.

Early work on pressure-cycled ventilators by Straub and Meyer (*13*, *14*) and Pavlin and Kadosch (*15*) was based on modified wall-reattachment fluidic amplifiers of great simplicity. Figure 3.10 shows the arrangement described by Straub. It consists of a wall-reattachment element with one output supplying breathing gas to the patient's lungs and the other output open to atmosphere. When the supply jet is attached to the left wall the inspiration phase commences and lung pressure rises. Feedback of this pressure ensures that, at a predetermined pressure, the jet switches to the vented output and allows exhalation. The feedback pressure then decreases until asymmetric entrainment characteristics, resulting from the set screw positions, causes the jet to switch back to the lungs. The ventilator also operates in the assister mode, since any attempt at inspiration reduces the feedback pressure more rapidly and thereby brings forward the commencement of lung pressurization. Flow rates, cycling pressures and frequencies are all controlled by preset variation of the feedback and bias control nozzle resistances. The figure shows typical data obtained with this unit for the lung deflexion/force ratio (compliance) usually found in human patients.

The work of Pavlin and Kadosch was similar to that of Straub but element geometry was chosen for maximum entrainment efficiency in order to reduce gas consumption as far as possible. However, the unit shows a ratio of expiration/inspiration time less than unity, which would be unacceptable in practice, and has no provision for varying the trigger pressure in changing from expiration to inspiration. For a given patient the airway resistance and lung compliance are fixed so the only possible adjustment is to the supply pressure and flow.

Although pressure-cycled ventilators are relatively simple they all suffer from the disadvantage of being sensitive to small leaks and are load-sensitive to changes in the lung characteristics. Ball (*12*) has carried out tests on a configuration, similar to Straub's, in which the relative values of control and output channel resistance, wall setback and angle, splitter distance, and aspect ratio were all varied. Their effect on load switching of the element at the completion of inspiration and the suction switching at the end of the expiration phase was investigated. Ball concluded that, for a fixed geometry, the optimum procedure was to remove the feedback path and to control the peak inspiratory pressure (and hence the tidal volume) by varying the supply pressure and control the expiratory pressure by varying the vented output restriction. The ratio of expiration/inspiration time was kept constant at 1·5, although it could be changed by varying the control port resistances at the expense of changing the switching pressures. In order to cover the need for cycling rate variation over a range of patient conditions, Ball has constructed a series of similar devices with different aspect ratios to provide the necessary flow rates. Figure 3.11 shows the typical inspiration-pressure profiles obtained with two different lung compliances.

It is clear that, with reservations, both pressure- and volume-cycled fluidic units can be used on patients. Ball has reported clinical tests with his pressure-cycled unit, both for minor surgery and in an intensive-care unit. Although air consumption was high, patients had been successfully ventilated for up to 24 h with the unit, and had also been able spontaneously to trigger inspiration themselves (assister-controller mode).

More sophisticated patient control is possible with the time-cycled ventilator at the expense of increased control circuit complexity. Vince and Brown (*16*) have described some preliminary work

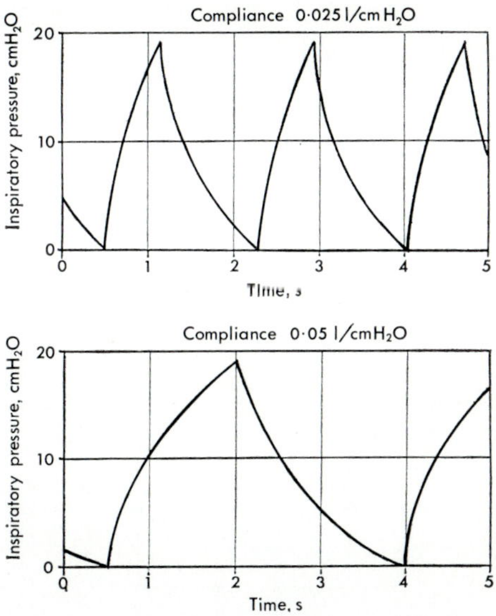

Figure 3.11 Variation of inspiration pressure with a pressure-cycled ventilator, after Ball (*12*).

on a fluidic system in which the inspiration and expiration times and the flow rate can be independently varied. The system is shown schematically in fig. 3.12. Timing is by means of a standard frequency oscillator operating at 12 Hz in conjunction with a series of binary counting stages to display the total number of pulses from the oscillator. This number is continually compared with a manually-set desired time interval. When the numbers coincide air is diverted either into or out of the patient. The range of inspiration and expiration times is 0·5 to 5 s, peak flow is 50 litre/min, peak pressure 70 cmH_2O, and the total air consumption 75 litre/min: 22 wall-reattachment elements are used.

A more practical circuit, again designed by Ball (*12*), gives variation of the expiration/inspiration ratio by pulse-width modulation techniques, in conjuction with a novel low-frequency oscillator, which dramatically reduce the number of fluidic components re-

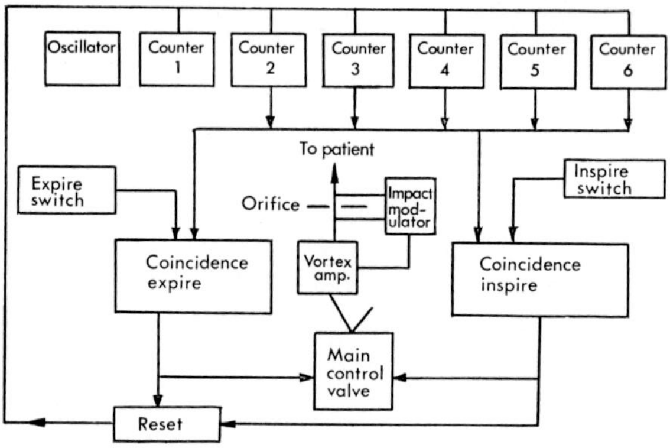

Figure 3.12 Logic diagram for time-cycled ventilator (*16*).

quired. The low-frequency oscillator circuit, which was originally patented by Brown (*17*), is shown in fig. 3.13. It consists of three bistable wall-reattachment elements *A*, *B*, *C* and two closed variable

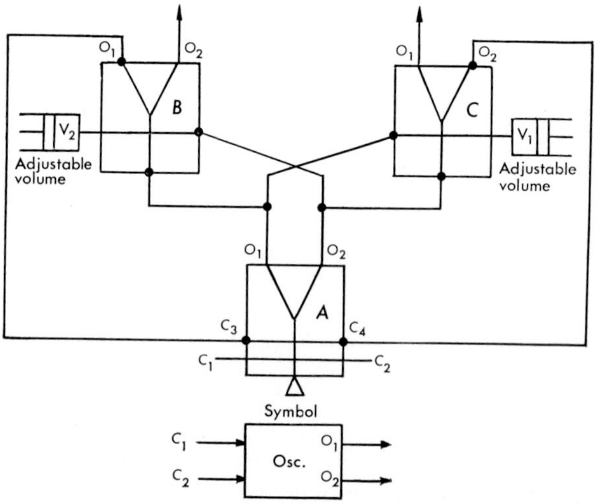

Figure 3.13 Low-frequency oscillator circuit, after Brown (*17*).

volumes V_1 and V_2. The capacitors cause vacuum switching of the supply jet onto their adjacent wall owing to the reduction in capacitor pressure by entrainment of air into the main jet. Assuming that the output of *A* is at O_2, element *C* has an active supply which initially passes out through O_1 as V_1 was filled in the previous part of the cycle. Also the same output from A charges V_2 in the inactive element B. The pressure in C_1 eventually drops because entrainment is followed by vacuum switching of the element giving an output at O_2. This signal is fed back to element A which switches to output O_1, thereby making element B active instead of C. A similar action takes place for the other part of the cycle, the duration now being dependent on capacitor volume V_2 instead of V_1. By varying the

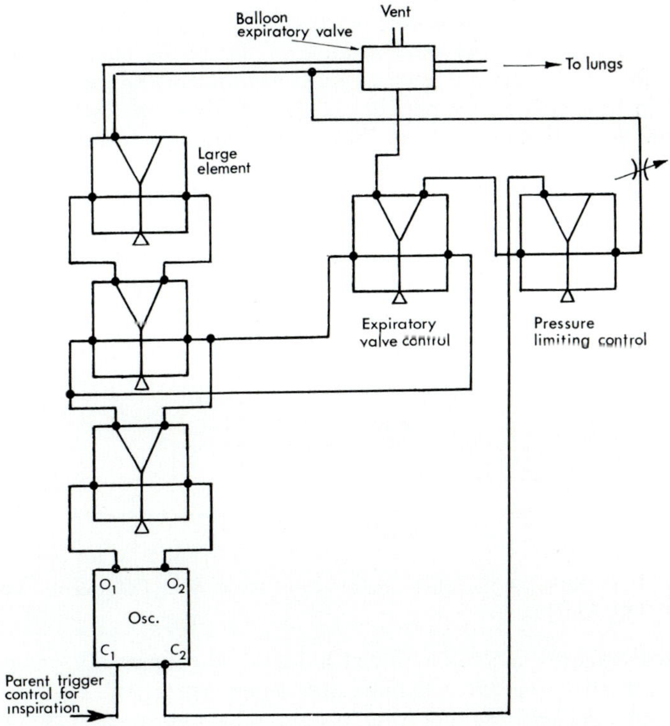

Figure 3.14 Time-cycled ventilator, after Ball (*12*).

capacitor volumes Ball found that mark/space time ratios up to 4 : 1 were possible with the configuration tested.

The basic ventilator circuit used by Ball is shown in fig. 3.14. It is similar to the volume-cycled unit circuit except that the buffer amplifier and large-scale ventilator element are switched from the oscillator circuit instead of pressure switches and the ventilator output passes directly into the lungs rather than into a bellows unit. The oscillator may be overridden by a pressure-limiting control if the inspiration pressure is too high, and also if the patient triggers the start of inspiration voluntarily (not shown in figure). A further refinement incorporated in a later design allows the inspiration- and expiration-time capacitor volumes to be changed during operation without malfunction of the circuit. The overall circuit contains virtually no moving parts and should be comparatively cheap to produce commercially by using integrated circuit techniques.

3.3 Further possible applications

In recent years considerable interest has been focused on artificial kidney systems for the maintenance of life in the case of chronic renal failure. Successful systems have been produced and are being used in hospitals or by patients in their homes (*18*). Some require pumps for circulating both blood and dialysis solution, while others have been designed for low flow resistance and can be used without pumps. All systems built at the time of writing are expensive to operate, despite successive reductions in the capital cost of equipment. Obviously the trend will be towards more automatic equipment, provided that it is both economic and reliable.

A paper by Montgomerie and Floyd (*19*) describes a fluidic automatic self-cleaning filter which may well find useful application in

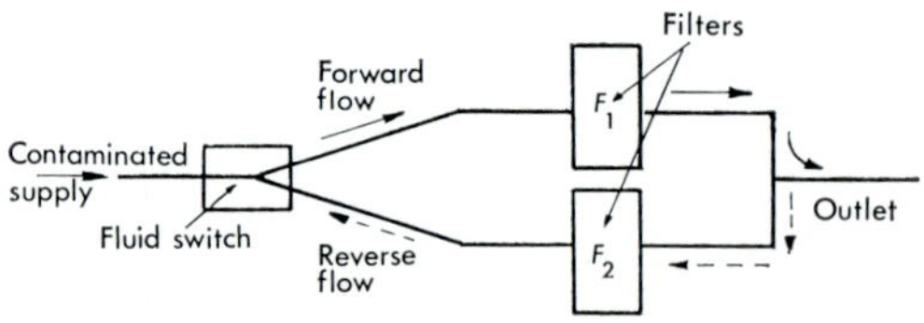

Figure 3.15 Self-cleaning filter circuit (from *Biomedical Engineering*, Vol. 2, No 10, Oct. 1967).

medical systems such as artificial kidneys. The simplicity of operation and the inherent reliability are again attractive. Figure 3.15 illustrates the self-cleaning filter system, which consists of two filter elements F_1 and F_2 connected to the two output channels of a wall-reattachment device. Consider the wall-reattachment element supplying fluid, usually a liquid, through filter F_1. Downstream of the filters the channels rejoin so that a back flow is established in the opposite limb of the loop. In this way filter F_1 removes dirt from the fluid whilst the reverse flow through filter F_2 cleans accumulated dirt from its surface. When filter F_1 has reached a certain level of contamination, pressure in the output channel of the switch is sufficiently high to cause the power jet to switch to the other side. This automatically reverses the rotational flow in the loop, making F_2 the active filter and F_1 the one that is cleaned.

Figure 3.16 illustrates the channel and filter housing configurations in more detail. Note that the filter housing is shaped so that, in reverse-flow filter-cleaning conditions, particles removed from the filter drop down through baffle plates into a settling chamber. This

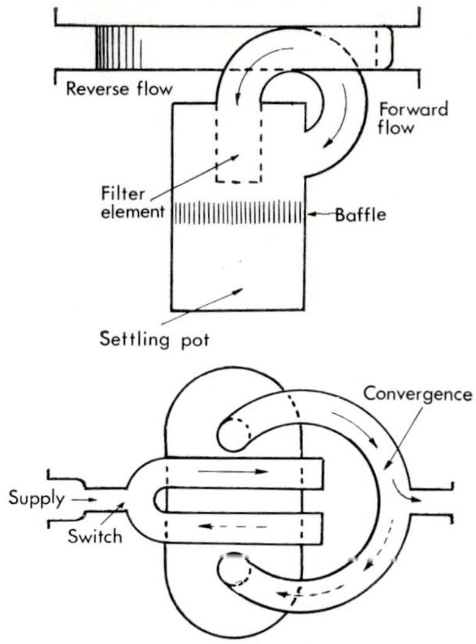

Figure 3.16 Details of filter shown in fig. 3.15 (*ibid.*).

ensures that when forward flow is re-established no old contamination is circulated in the system again.

With suitable design the switch can be made to operate consistently at a predetermined value of filter pressure-drop, so that the whole system operates to keep the pressure-drop fixed and low, irrespective of the contamination load. Figure 3.17 compares the

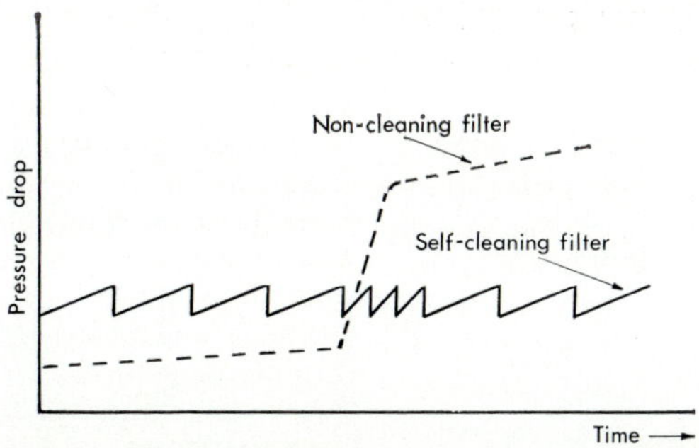

Figure 3.17 Effect of varying contamination rate on characteristics of filter shown in fig. 3.15 (*ibid.*).

characteristics of the new filter with conventional filter elements for fluctuating contamination.

It is often necessary to control fluid in other ways than by simple switching and diversion. For instance, in the ventilator circuits described earlier it is often necessary to fit an expiration valve which allows full flow in one direction and shuts off completely to prevent

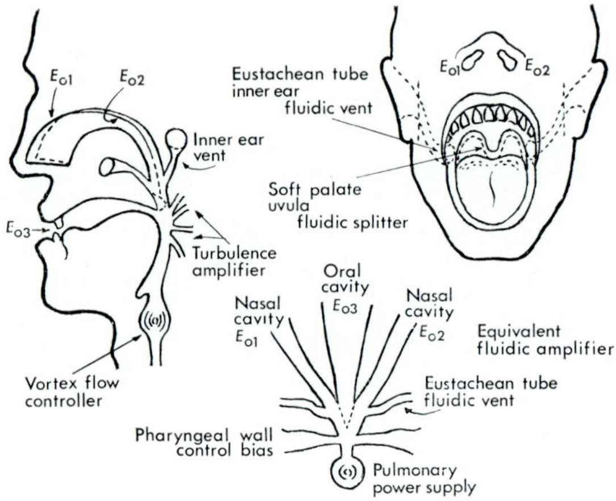

Figure 3.18 Fluidic model of oral system, after Petrovick (*20*).

reverse flow. Ball has described a simple inflatable balloon valve for this purpose, and pure fluid devices with approximately similar characteristics (such as a vortex diode) can be used as an alternative. Petrovick (*20*) has carried this philosophy a stage further and has modelled the whole of the human oral cavity dynamics in terms of equivalent fluidic devices, as shown in fig. 3.18. The equivalent relationships suggested are:

HUMAN	FLUIDIC
Pulmonary system	Pulsed pressure power supply
Vocal chords	Vortex flow modulator
Pharyngeal wall and salpingo	Turbulence amplifier
Upper pharyngeal cavity	Fluidic diode
Soft palate	Variable fluidic diode
Eustachian auditory tube	Atmosphere vent
Uvula	Amplifier splitter
Nasal septum	Amplifier interaction region
Nasal cavities	Outputs E_{01} and E_{02}
Oral cavity	Output E_{03} which has capacity
Nasal turbinates	Flow restrictors
Lips	Variable-output-impedance valve
Tongue	Variable fluid capacitor

Preliminary results presented by Petrovick are encouraging enough to suggest that fluidic modelling may give better insight into oral physiology. Presumably the same considerations may also be extended to the modelling of other areas of the human system.

There is a continuing quest for improvements in a wide variety of aids for the disabled. Many such electrical, pneumatic and mechanical systems already exist, some of which are very sophisticated. With the expansion of technical knowledge concerning fluidic devices and systems it is possible that applications to some of these aids could be usefully made.

For instance, some attempts have been made to simulate muscles using bundles of flexible pneumatic tubes. If such devices became practical they would be very compatible with fluidic systems. One can visualize switching flow into and out of different combinations of tube bundles to obtain a range of human limb characteristics.

To take another example, there are many paralysed patients who can only carry out simple tasks for themselves. Various tactile, blowing and sucking devices are possible, using fluidic sensors controlled by means of pneumatic fluidic systems to provide a range of command functions such as closing doors, switching off lights, emergency bells, etc.

Cushions of air are being used increasingly for treatment of severe burns. This requires careful, reliable regulation of the airflow round the patient. It is possible to make fluidic circuits to achieve this type of control, and at the same time to provide additional facilities for safeguarding the patient.

Although there is no published work on fluidic applications to automatic analysis equipment for, say, blood and urine sampling, undoubtedly considerable interest has been developed in this area. The problems are essentially similar to those of mechanical handling associated with pneumatic and mechanical systems in many industrial situations. It therefore seems likely that equipment developed will be cheaper and more reliable than existing units.

4

INDUSTRIAL APPLICATIONS

G. M. Brewin

Mullard Southampton [*Formerly with Techne* (*Cambridge*) *Ltd*]

The basic ideas used in the developing technology of fluidics have been known for many years and it is now over ten years since these ideas were first used collectively to form a fluidic control system. There have, however, been very few large-scale applications of fluidics in general industry to date, and by examining the basic reasons for this it is possible to locate the various fields where fluidics have an important part to play. Admittedly there are on the market industrial products which utilize fluidic techniques, but in most cases performance depends on the action of a few fluidic devices, often only one, and the result can hardly be described as a fluidic control system.

The real market for fluidics in industry ought to be in the field of machine control, yet very few instances have been reported of fluidics in use. Of course, fluidic devices can be incorporated in a control circuit and made to perform control functions previously carried out by electronic devices, but such substitutions do not on their own show that there is a general field of application available. They merely show that something is possible if an effort is made to achieve it. If there is a place for fluidics in machine control it will be demonstrated only if fairly large numbers of essentially similar control units are used regularly by industrial organizations.

4.1 Lag in application

To date there have been few examples of industrial application to machine control. In the period during which electronics has progressed from the miniature thermionic valve *via* the transistor to completely integrated solid-state logic systems, fluidics has barely moved from the laboratory to the factory floor. The reasons for this are not to be found by studying the basic logic devices but by looking at the ancillary devices that are required for a complete control system.

Over several decades the electronics industry has developed a very wide range of sensors to suit every requirement of the control engineer. In general, as the design of control systems has developed from valve-based to solid-state electronics, the range of 'sensor interfaces' has been sufficient to meet the designer's needs. The development of new control techniques has therefore not been delayed whilst interface devices were being developed. This has not been the case with fluidics.

The basic advantages of air valves with no moving parts have attracted a large number of research workers. In many instances they have duplicated the work of others on basic logic elements whilst the relatively simple, but nevertheless necessary, design of interface units was neglected. The result has been that the control engineer, considering the possibility of using a fluidic control system, has found a very limited range of interface devices for fluidic circuits but an almost infinite range for use in electronic systems. There is little wonder that, without very marked advantages to be gained by using fluidics, electronics has been the choice of the majority.

Fluidics is similarly situated in relation to conventional pneumatics or hydraulics. Conventional valves are often considerably cheaper than fluidic devices. Yet, although existing trip valves and other comparable interface units can be used in a fluidic system, they are rather clumsy when compared with the ideal fluidic interface unit, which can be readily visualized but is not available in large quantities.

4.2 The definite advantages

Fluidics must, therefore, have very definite advantages in a particular situation before it will be considered seriously. For general industrial applications the advantages come under four headings: high reliability of the interface unit, simple construction of the logic circuit itself, inherently flameproof units, and possible cost advantages in the near future.

Reliability. Most electrical interface units use some form of electrical contact to translate mechanical movement into an electrical signal. Failure of the unit is either mechanical or because of contact deterioration. Although there are several methods for overcoming the relevant defects the results are nearly always more expense and complication. Using fluidic techniques one can eliminate the mechanical movement of the switch and the contact operation can be dispensed with completely. The result is much higher reliability. Also it is very difficult to make an electrical switch unit with no overtravel, unless the actual parts of the controlled machine do the switching. With a fluidic back-pressure sensor the rate of airflow through the switch can be adjusted so that the switching point is at the limit of machine movement. This can lead to greater accuracy in a machine's performance.

Construction. Pneumatic logic circuits are much easier to construct than equivalent electronic circuits, both at the prototype stage and in general production. This is because interconnecting pipework is easily installed and is not subject to the 'dry joints' that can cause trouble with electrical wiring. Not only are fluidic devices easier to assemble but they are also more readily understood by works maintenance staff, once the basic details have been assimilated. This is particularly an asset in small- to medium-size organizations, where maintenance work on electrical controls requires the attention of a visiting service engineer, with the usual delays before the system is working again.

Flameproofness. Fluidic control circuits are of course inherently flameproof. This virtue can carry a great deal of weight when controls are considered for chemical plant involving inflammable substances. Control engineers have not ventured far into this field yet, but there is considerable scope for linkage of fluidic methods with existing plant control techniques using air at 3–15 lbf/in^2 (20·7–103·4 kN/m^2) pressure as the control medium. Occasionally doubts have been expressed relating to the possible build-up of a static electrical charge on equipment by air flowing to exhaust, but there has not so far been any report of this happening in practice. I am very doubtful whether it would occur at all, particularly as the speed of the exhausting air is quite low.

Cost. At the moment it is often more expensive to construct a fluidic control unit from commercial devices than to construct a conventional electronic one. However, if the basic methods used in manufacturing fluidic units are examined it becomes obvious that the present-day price bears little relation to manufacturing cost. At present the major part of the cost of elements covers the vast effort in technical services that is needed to prepare the market for the sale of fluidic devices in quantity. As sales rise the cost of fluidic elements should fall rapidly, and the reduction will most probably be much greater than can be achieved for either electronic devices or conventional pneumatic valves. In a few years' time, therefore, it should be possible to construct a fluidic control circuit for considerably less than a conventional electronic unit.

4.3 The limitation of speed

Fluidics will not be the universal answer to control problems that was first thought. The main problem is speed. Pneumatic signals cannot travel faster than the speed of sound, which is about 1000 ft/s (305 m/s), so there is a limit on the distance over which control can be exercised by fluidic elements. The range depends on the overall speed of response required, but in practice 100 ft (30 m) seems to be the reasonable limit of signal transmission at the time of writing.

Fluidic elements take much longer to switch than electronic devices do, though they are much faster than conventional pneumatic valves. The average switching time of present-day units is about 1 ms, to which must be added the delay caused by both the length and volume of the interconnecting passageways.

The slowness of fluidic devices, when compared with their electronic counterparts, places an upper limit on the possible complexity of control systems using fluidic techniques.

It seems likely that anything beyond the 'point-to-point' numerically controlled machine will have to be electronically controlled in order to have a speed of operation which is commercially acceptable. But in a vast number of cases the overall speed of operation of a machine is limited by the inertia of its moving parts. As long as the speed of a fluidic control system is greater than this, the limit on switching speed will not have any adverse effect on the performance of the whole unit. Slower switching can be an advantage. Spurious signals that last long enough to affect an electronic system are often far too short to have any effect on a fluidic unit.

4.4 Disadvantages of cost and size

The two other disadvantages of using fluidics are the inherently higher cost of compressed air and the rather bulky connexions that are needed between individual units.

Mechanical problems associated with connexions between elements do put an upper limit on the size of a fluidic control unit, but in industrial applications where a reasonable number of similar systems are to be constructed the cost of an integrated system of interconnexions is easily justified.

The extra cost of running a control system on air as opposed to electricity has to be balanced against the advantages. It should be mentioned that fluidic elements usually run at a higher power level than electronic devices, and often the output of a fluidic element need not be amplified before being able to do useful work.

4.5 Three categories of application

Taking these various advantages and disadvantages into account, industry in general is beginning to use fluidics in three main categories; jet sensing, programmable sequence control, and flameproof equipment. The use of an air jet to sense or detect an object without solid contact is a simple alternative to photo-electric and ultrasonic methods, and avoids necessity for complex electronic components. For programmable sequence control, fluidics offers the ability to read punched cards directly using air streams. This is now considered preferable to conventional electrical plugboard and uniselector switching. As previously mentioned, the use of fluidics for flameproof equipment is not far advanced, but from the few examples at present in operation a major expansion is expected in this application in the near future.

4.6 Sewing-machine control

One of the best examples of fluidic jet-sensing techniques is the system designed by the Marine Engineering Co. of Stockport for industrial sewing machines (fig. 4.1). Air-jet sensors are used to detect the passage of the garment to be stitched and they enable the operator to exercise much closer control over the machine.

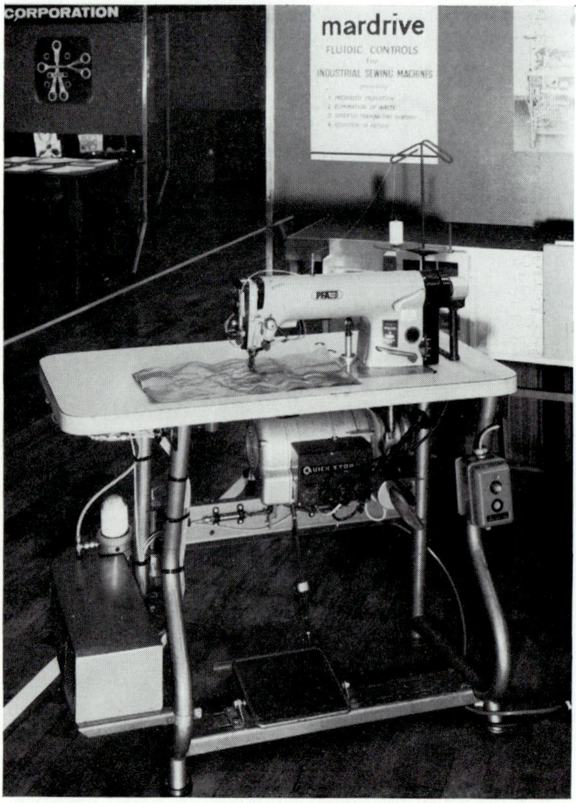

Figure 4.1 Lockstitch sewing machine with fluidic control.

The sequence of events effected by the fluidic system in a lock-stitch machine is straightforward and fairly representative of the sequences in a variety of stitching machines. Essentially the control system does not take over from the operator but governs the start and particularly the completion of stitching so that wastage is reduced and production increased.

The sequence is as follows. The operator presents a garment to the machine and in so doing breaks a low-pressure air-jet. This causes the presser foot to fall and the operator's treadle control to be linked to the driving motor. The machine is then completely under the operator's control. She can stop and start or vary the speed of the machine. The machine remains in this mode of operation until the trailing edge of the garment passes through the air-jet. Re-establishment of the air-jet causes the machine to go through a finishing cycle which prevents the stitching from overrunning the edge of the garment. The steps in the finishing cycle are: (1) detect trailing edge of garment, (2) reverse feed (backstitch), (3) return feed to forward direction, (4) disengage driving clutch, (5) stop needle at top of its stroke, (6) cut and trap both top and bottom threads, (7) lift presser foot to allow removal of finished garment. This sequence of events could be completed by the operator but the use of an automatic system ensures consistency and reduces wastage. In addition it removes a great deal of the skill from the operation and therefore makes operator training much simpler.

The sequence could of course be controlled by means of an electronic system and photo-electric sensing devices, but in the garment industry the servicing of such a system could be difficult as sewing-machine mechanics are not trained to look after this type of equipment. Fluidic equipment has been readily accepted, and the necessary maintenance training can be completed in a very short period.

4.7 Machine-tool control

The success of this control unit lies in its ability to operate from the signals developed by air-jets spanning a distance of over 1 in (2·5 cm). For this purpose the turbulence amplifier, operating at a very low control pressure, is at a distinct advantage over other forms of fluidic element. In programmed sequence control and flameproof equipment the turbulence amplifier does not have so many advantages, and hybrid systems are often found. Such systems combine several types of fluidic device, and high-pressure pneumatic units are often used as well. An example is a fluidic system for controlling capstan lathes, developed by Techne (Cambridge) Ltd, which at the time of writing is in full production.

The Techne system uses almost every pneumatic technique available to the control engineer and is not merely fluidic. This is because the main need with any such product is for reliability at as low a cost as possible. Although fluidic techniques such as air-jet sensing could be used to detect the movements of machine slides, the necessary pressure-amplifying valves are expensive and it is cheaper to use conventional pneumatic trip valves. Especially is this so because the output of the valves is required to operate further valves using high-pressure air.

The programming unit is the heart of any programmed control system. Fluidics offers a wide variety of ways in which program

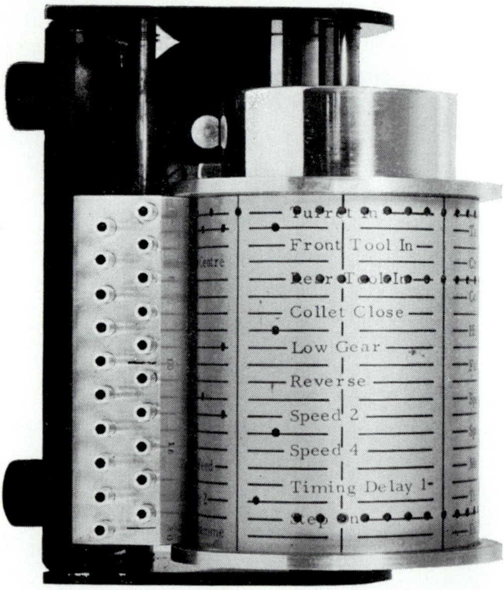

Figure 4.2 Punched-card program on pneumatic reading unit.

information can be stored and then read by the machine. The conventional electronic method of the plug-board can be used, but a punched card is far more satisfactory for carrying program information. Low-pressure-air sensing of punched cards is probably the simplest method known for reading this type of information, and there should be increased interest in it in the near future. Using air it is possible to read information at speeds up to 4 kHz, which is more than adequate for most industrial applications, and the reliability is much greater than that of electrical contact systems.

In the Techne capstan–lathe system the punched card containing the program information is wrapped around an accurately machined drum as shown in fig. 4.2. Rotation of the drum then causes the

Figure 4.3 Control circuit for capstan lathe.

punched holes to move past a reading head which detects them by means of a back-pressure air-sensing system. This method—moving the program past the reading head—is the simplest for reading a punched-card program.

Figure 4.4 Capstan lathe with fluidic programmed sequence-control unit.

The alternative is to switch the 'reading' air system from row to row, using either a mechanical valve or a pneumatic ring counter. The pneumatic valve has the same limitations as the moving program system, for mechanical movement is required to return from the end of the program to the beginning. The ring counter is the obvious method where the speed of operation is high. However, for the equivalent of an electrical plug-board of say 25 rows and 30 steps, the ring counter contains a large number of fluidic elements and consumes a considerable amount of air.

The problem of returning a program to the beginning, from the end of a cycle of events which is shorter than the complete circumference of the drum would allow, is economically overcome in the Techne lathe control unit (fig. 4.3). The function is performed during one of the machining operations when several seconds are available in which a machine (like that shown in fig. 4.4) does not need to read further instructions. A special programming technique is involved but it is easily learnt by the machine operator and does provide a simple and cheap solution to the problem.

4.7.1 *Features of fluidic circuits*

Machine movements are controlled in a fairly conventional way using pneumatic and hydraulic cylinder and piston arrangements, and I do not intend to describe them. However, the fluidic control circuit has a number of interesting features which are applicable to all types of fluidic system used in industrial situations, and these are worth dwelling upon.

One frequent problem is the loss of information from memory units in the control system when the power supply is turned off, either on shut-down or on failure of the supply equipment. This problem is of course not new to designers of electronic systems, but mechanical engineers who are used to conventional valve systems can find that they are faced with the problem after a system has been constructed. Considerable effort is then required to rectify the matter.

In a conventional pneumatic valve system the friction between the valve spools and associated seals is usually sufficient to retain the spool in its last position if the air supply to the system is removed. In a fluidic circuit, on the other hand, memory elements can often start up in either mode. It is important to ensure that position on start-up is consistent, particularly when information contained in the memory controls the position of an expensive machine tool.

There are two main methods for doing this automatically. One is to incorporate a bias in the memory. The other is to provide a special 'clamp' signal during start-up. The bias method has the advantage of simplicity, but practice has shown that it is not always consistent. If the bias is pronounced to ensure maximum consistency, then the switching of the memory unit is affected during normal circuit operation. The 'clamp' method has been shown to be far superior in this respect. Figure 4.5 shows some simple arrangements for ensuring consistency during the start-up and shut-down of a system.

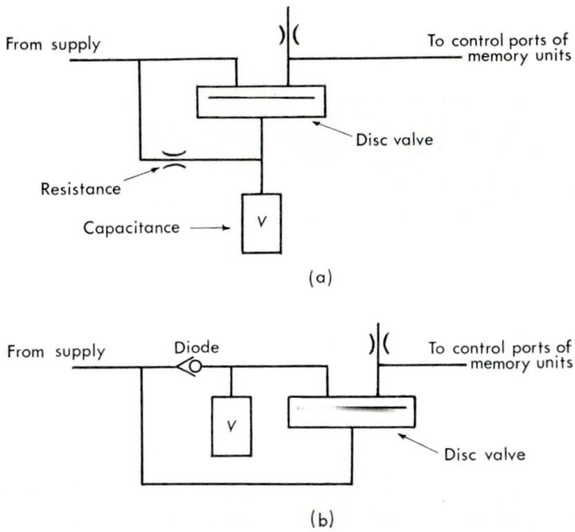

Figure 4.5 How consistency is ensured in start-up and shut-down. (a) Start-up control. Disc valve conducts until pressure in capacitance reaches supply pressure. Output for 2–3s is usually sufficient. (b) Shut-down control. Disc valve conducts if supply pressure falls below certain level determined by part configuration. 70–80% is usually satisfactory.

Shut-down is just as important as start-up. Quite often a fluidic system will perform in a very satisfactory manner until the air supply is turned off. Then all manner of things happen. In industrial applications, where small control units often control many thousands of pounds' worth of precision equipment, this is a point which requires particular attention.

4.8 Process-plant control

To date the major effort in applying fluidics to industrial control has been in the field of machine tools, but it is not expected that this will remain the major field of activity for very long. The control of process plant appears to offer far wider scope for the application of fluidics. In this area the high reliability and inherent flameproof qualities of fluidic devices far outweigh their slow response and fairly high air consumption.

Figure 4.6 shows a flameproof programmed sequence-control unit. It was constructed by Techne (Cambridge) Ltd for an organic peroxide plant manufactured by Simon Carves Chemical Engineering Ltd. This system marks the departure from earlier fluidic systems using individual elements interconnected with flexible tubing. The

Figure 4.6 Programmed sequence-control unit for chemical plant.

future systems which it presages will have completely integrated circuits.

In this system approximately half the elements have been integrated into a machined aluminium block in order to reduce the number of interconnexions used and increase the reliability of the system. As systems become more complicated some degree of integration is essential to provide a system that can be managed by average maintenance staff. Almost certainly the reason for the slow application of fluidics to the control of complex process plant is the delay in transition from interconnected elements to integrated circuits. This must be the main area for effort in the next few years.

NOTE: I acknowledge the help of the Marine Engineering Co. (Stockport) Ltd, Techne (Cambridge) Ltd, and Simon Carves Chemical Engineering Ltd in preparing this chapter.

5

NUCLEAR APPLICATIONS

M. J. Fisher
British Hydromechanics Research Association

The high reliability of fluidic devices and their ability to operate in high temperature and radioactive environments makes them particularly suitable for instrumentation and control equipment for nuclear plant.

The coolant temperatures of gas-cooled reactors already exceed 450°C and are likely to reach 650°C for both advanced gas-cooled and fast reactors. The high levels of temperature and radioactivity limit the choice of materials that can be used. This severely restricts the design of any moving-part mechanical equipment, particularly that using sliding surfaces. Only the simplest electrical equipment, such as thermocouples and electromagnetic pumps, can operate in a reactor core. Much of the conventional equipment is usually outside the core, but the reactor vessel has to be breached to connect the equipment to the core, which is very costly. Fluidic systems can be in the core itself. They do not make stringent demands on materials. Fluid supply is not a problem because the reactor coolant itself, whether liquid or gas, can be used as the operating medium.

Safety is another important aspect of reactor design. Systems are used to monitor the operation of a reactor and its ancillaries so that it is shut down if a fault is detected. It is normal practice not to rely on a single system but to have several systems operating in parallel. Most of these would be electrical systems. Fluidics can be used with the advantage that the failure modes are quite different from those of electrical systems and that all of the systems are therefore unlikely to fail at the same time. A fluidic system depends on the electrical drive to a compressor or blower, but a standby supply of compressed gas can be provided. Like reasoning applies to conventional pneumatic systems, but fluidic systems have the advantage of greater reliability, which is at a premium here. Also they have greater flexibility in that they can be placed, in part or in whole, in higher-temperature and radioactive locations.

Another hazard which must be guarded against in reactor design is damage by burst pipes, particularly those carrying reactor coolant, for its escape could cause local overheating in the core. A non-return valve placed in the line would reduce the coolant loss from a burst upstream of it. The absolute shutting-off ability of a moving-part non-return valve would be ideal here. However, as a safety device the valve should very rarely be required to operate. It might work only once, if at all, in the life of the reactor. It is very doubtful whether any moving-part device would have adequate reliability in these circumstances. Although fluid diodes—devices with a larger resistance to flow in one direction than in the other—do not have absolute shutting-off ability, the absence of moving parts gives them the reliability needed. They cannot contain the fluid completely, but at least they reduce the rate of loss.

One difficulty in presenting a coherent picture of work in the nuclear field is commercial security. This seems to be more stringent than defence security in covering not only how things are done but also what is being done. It is known that two of the larger U.S. companies with commercial nuclear interests are working on fluidics, but they have published little if any of their work. In this country the U.K.A.E.A. did publish some feasibility studies but nothing has been published since. The full picture of commercial developments may not become clear until the next generation of reactors goes into operation. Since there is usually a lead time of about five years between designing and completing a reactor, this may not be until the middle seventies.

5.1 Systems operating with reactor coolant

The examples chosen for description in this section are a system for locating burst fuel-elements, a temperature sensor and a quality meter. Diodes are discussed separately in the next section.

The system for detection of burst fuel-elements (*1*, *2*) makes use of wall-attachment elements to route samples of coolant from the outlet of each fuel channel to a radioactivity detector. A rise in the level of radioactivity indicates a burst. The samples are fed into the power nozzles of the elements, one of which is switched in one direction to feed its sample to the detector whilst the rest are switched in the other direction to return their samples to the reactor (fig. 5.1).

Wall-attachment devices are not perfect diverter valves. Depending on the differential pressure between the two outputs, the elements could induce or spill over some of the input flow. Working, for instance, with sixteen samples, if each element were spilling over 1% of its flow then the required sample would be diluted by 15% compared with the other samples. If the elements were arranged in four rings of four, each ring feeding another ring of four, the dilution would be reduced to 5%. A binary break-down, where appropriate, gives the minimum dilution.

The actual spillover or induced flow depends on many factors, such as temperature range, manufacturing tolerances, sample flow

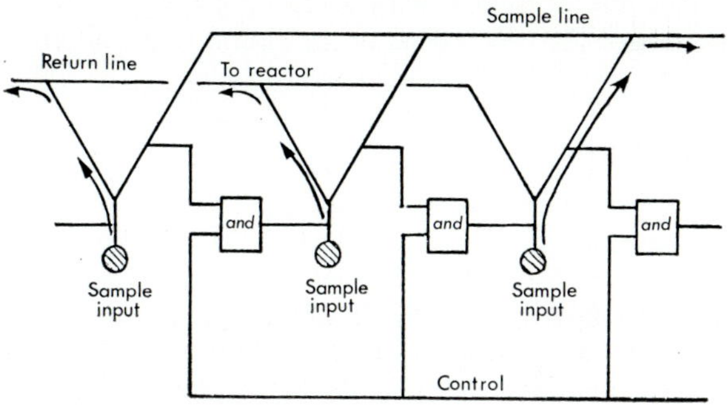

Figure 5.1 Burst fuel-element detection: sampling system (*2*).

rate, etc., and they determine the overall performance of the system. The elements are switched by connecting them in the form of a ring counter (fig. 5.2) using other fluidic elements to provide the ring counter logic. A ring counter form of control minimizes the number

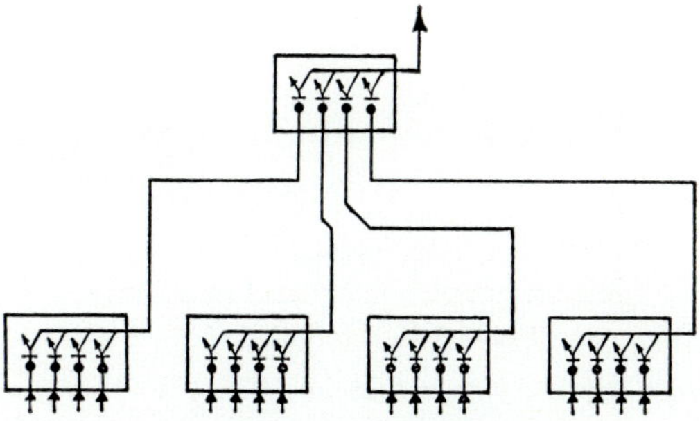

Figure 5.2 Staging of rings (*2*).

of control lines, and hence the number of times the reactor vessel has to be breached.

The temperature sensor was first devised for use in aerospace applications. It is now being developed for the measurement of metal vapour temperature. Sodium and potassium, probably used to drive nuclear reactor turbines, are the metals in mind (*3*).

The quality meter is based upon the same principle but is used to measure the 'quality' of a mixture of liquid metal and its vapour (*4*). 'Quality' here is the proportion of vapour in a liquid–vapour mixture and is normally assessed from the difference between the flow rate measurement of the mixture and that of the liquid only. Free gas in a liquid, even in small quantities, markedly reduces the velocity of sound because the bulk modulus of the mixture is considerably less than that of the liquid alone whilst the mean density is only slightly different. So, for a given temperature and pressure, the resonant frequency of a line decreases with increasing quality. As with the temperature sensor the line is used in the circuit of a wall attachment oscillator.

5.2 Diodes

Three types of diode have received attention for reactor applications.

Probably the best known and simplest of them is the *vortex diode*, described in Sections 1.8 and 2.4.2. The device is in its high-resistance state when fluid flows in through the tangential port, forms a vortex in the chamber and leaves through the axial port. With flow in the other direction the device is in its low-resistance state—a vortex is not formed.

The performance of diodes is usually quoted as the ratio of the pressure drops for the same flow in the two directions, though sometimes the flow ratio for the same pressure drop is quoted. The former is approximately equal to the square of the latter.

Although fluidics is thought of as a new technology the earliest

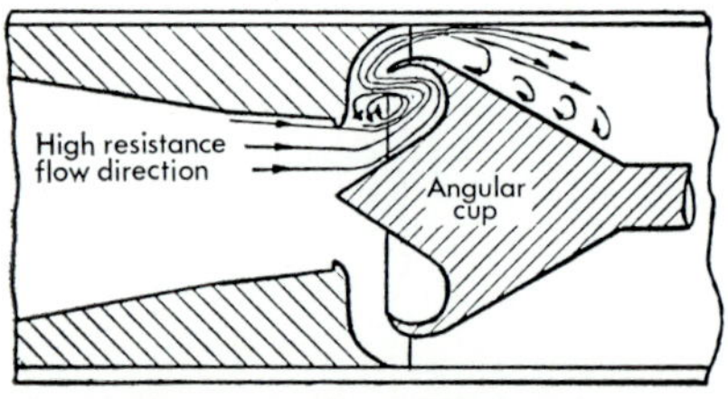

Figure 5.3 Scroll diode (*7*).

publication on diodes was made in 1920 (*5*). The first report on vortex diodes, made in 1935 (*6*), quoted pressure ratios of up to about 40 : 1. More recent work quotes ratios of 39 : 1 (*1*) and 50 : 1 (*7*). Higher ratios could be achieved but only by using excessively large vortex chambers.

The input and output lines are at right angles, which could be an advantage in some applications. In general parallel input and output lines are preferred, so a right-angled bend must be made in the pipework leading to the diode.

Cavitation affects performance. In the forward direction cavitation occurs in the tangential port, which increases resistance and reduces the effective pressure ratio. This can be important where the working pressure is close to the vapour pressure.

The *cup* or *scroll diode* (fig. 5.3) has a geometry more suitable for in-line pipe arrangement. Here a larger annular vortex is formed in the reverse direction. The performance is poor. Pressure ratios reported are about 8 : 1 (*1*) and 6·6 : 1 (*7*).

Another in-line device is known as the *fluid flow rectifier* or *cascade diode* (fig. 5.4). (The terms 'rectifier' and 'diode' illustrate that the antecedents of those who develop fluidics are as much in electrical as in mechanical engineering.) The device has many stationary blades, shaped like turbine blades, attached to a rod fixed along the axis of a tube. The blades are positioned so that if fluid enters from the left it is deflected and is caused to spin. After it has passed the first few

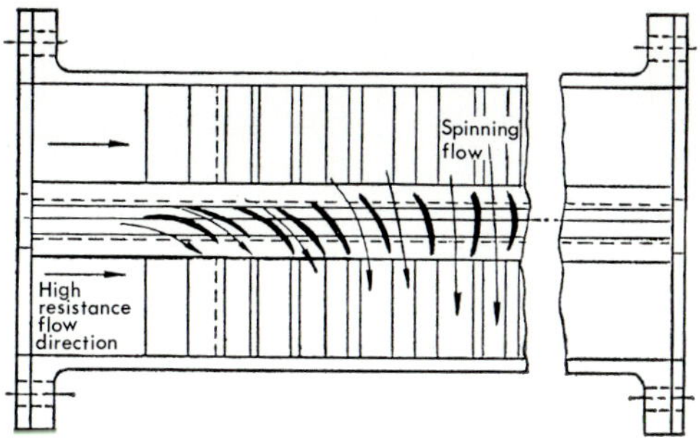

Figure 5.4 Cascade diode (*7*).

blades the flow velocity is almost at right angles to the tube axis. The vortex induces a large pressure loss, and the device is in its high-resistance state. Flow in the other direction meets the device in its low-resistance state, for the blades present a bluff obstacle. The fluid does not spin but passes almost axially down the tube between the rows of blades.

Pressure ratios as high as 181 : 1 have been reached (*8*) and higher ratios can be achieved.

Fluid flow rectifiers are less prone to cavitation than vortex diodes. In fact their performance is better than that of any other fluidic diode. But they are relatively complex, and their expense can only be justified in critical areas.

5.3 Neutron flux detectors

Two approaches to the problem of neutron flux detection have been reported, one by Bendix and the other by Rocketdyne.

The Bendix report covers only a feasibility study (*9*). The neutron flux is used to cause a pressure by ionizing a stream of gas in an electric field. This is fed to a vortex amplifier the output characteristic of which is used as an approximation to a logarithmic function.

In the Rocketdyne development (*10*) gas at constant pressure and temperature is fed *via* sonic orifices into two 'legs' (fig. 5.5) so that the flow is 'choked', that is independent of downstream conditions (cf. Section 7.3.4). The active leg contains material which absorbs both neutrons and γ-rays whilst the compensating leg absorbs only the γ-ray flux. Thus there is differential heating of the two legs. Differences follow in the temperature and in the pressures across the outlet

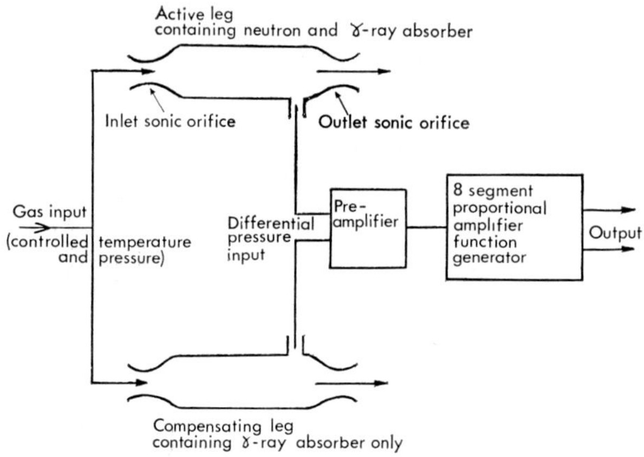

Figure 5.5 Rocketdyne's neutron flux detector.

sonic orifices. The pressure difference depends on the neutron flux and is amplified by proportional amplifiers. Connected in parallel the amplifiers form a function generator giving a piecewise approximation to a logarithmic characteristic. In 'bread-board' (i.e. experimental preliminary) work on the design the neutron flux was simulated by heaters. Corning amplifiers, modified to improve flow and pressure gain, were used in the function generator.

5.4 Research reactor applications

Sequence systems using wall-attachment devices have been developed for the Euratom reactor in Italy (*11*). They are for use in the irradiation of samples in the core of a reactor. One of them is a pneumatic system for moving samples in cylindrical capsules into and out of the core. This type of facility is descriptively named a pneumatic 'rabbit' system. The samples are stored in drum-shaped magazines which can be rotated to bring any one of a number of stations into line with the conveying tube. Pneumatic signals from drum stations and various valves are fed to the fluidic circuits which sequence the loading and unloading operations.

Another system moves a rod, with a sample attached to it, into and out of the core at a low frequency (typically one cycle in twenty seconds or 0·05 Hz). Binary counters divide the frequency of a fluidic variable oscillator to give output timing pulses in the range of 10 to 40 s. The frequency is measured by comparison with a fluidic reference oscillator. The pressures in the cylinder driving the rod are sensed by Schmitt triggers which operate conventional on/off valves to control traversing speed. Some difficulty was experienced with reference oscillator stability, and the comment was made that perhaps a tuning-fork oscillator would overcome this problem.

5.5 Future prospects

Fluidic technique for out-of-core use is likely to follow the mainstream of fluidic development. Problems being encountered at the time of writing, such as the effect of oil and dirt in the air supply, will be less serious outside the core because cost is not the major factor and the expense of producing a clean air supply can be tolerated.

For in-core use the technique developed will have little in common with normal fluidics, since the problems are specific to the nuclear industry. A perennial problem is the measurement of variables for a large number of fuel elements. Fluidic multiplexing, to limit reactor vessel penetrations as in the burst fuel-element detection system, is an attractive proposition. This type of system could be applied to the measurement of other variables. An ingenious scheme has already been suggested for a blocked element detector (*1*). In diodes fluidics has already been applied to operating on the main coolant flow. Perhaps other fluidic devices could be developed for controlling the coolant.

6

MARINE APPLICATIONS

M. J. Fisher
British Hydromechanics Research Association

Fluidic systems for marine application have been under development for only the past few years, the earliest report of any work in this area having been made in 1964. Most of them are in the prototype stage, only a few having reached seagoing trial. None is yet in use as standard equipment.

The developments are mainly being carried out for military applications. They cover a varied field, exploiting fluidics in many forms. Digital devices find a role for sequence systems, proportional devices for servo-controls and large-scale devices operating with water. Fluidic devices are competing as much with conventional pneumatic control gear and large moving-part valves as with electronic equipment. The main reason for using fluidics is the prospect of high reliability, a major factor in the choice of all marine equipment.

6.1 Sequence systems

A typical example is the prototype automatic starting and protection system for standby diesel generating sets developed by Plessey in conjunction with the Admiralty Engineering Laboratory (*1*). The protection system, which monitors engine variables during normal running, is separated from the automatic starting equipment so that the latter can be shut off to minimize air consumption once normal running speed is reached.

After the start-up signal is given by a manual push-button the system checks that the engine is in a suitable condition for starting. This includes checking that manually operated valves in both the distilled- and sea-water sections of the cooling system are in their correct position, that the levels of fuel and distilled water in the tanks are above a certain value and that adequate air pressure is available for the starter motor. The sea water and air-hydro lubricating oil pumps are then primed. After 15 s the oil pressure is checked to ensure that it has risen above 0·07 N/mm² (10 lbf/in²). Then reduced air pressure, fed to the starter motor to turn the engine slowly for two revolutions, is followed by the full air pressure required for starting. If the engine fails to reach 800 rev/min in 10 s the air supply is shut off for 15 s and a further 10 s attempt is made to start the engine.

If at any point in the starting sequence the necessary conditions are not satisfied the sequence is stopped and the appropriate fault indicated. For example, a closed sea-water inlet valve would stop the sequence before the priming of the pumps.

Two control cabinets are used, one local and another remote from the engine. The local cabinet contains the fluid logic and has separate indicators to identify individual fault conditions. The remote cabinet gives a simplified fault indication. The sequence may be started from either cabinet.

The engine protection system monitors various operating variables, e.g. cooling-water temperature, some of them at two levels. Faults are classified as of low priority or high priority. If a fault is of low priority an alarm is sounded and the fault identified but the engine is kept running. With a high-priority fault the engine is also shut off. An example is the cooling-water temperature, which has a low-priority level of 75°C and a high-priority level of 80°C.

The logic circuits use wall-attachment *or-nor* devices. Timing signals in 5 s multiples are obtained from an *or-nor* resistance–capacitance feedback oscillator giving an output of one cycle in 2·5 s (0·4 Hz) followed by a shift-register stage dividing by two. The 5 s signals are fed to a ring counter, again using shift-register stages to give the various time intervals required in the starting sequence. To overcome problems arising from spurious signals when the air supply is switched on, passive momentum-interaction *and* gates are included in the shift-register stages. One input is fed from a pressure derived from the supply so that signals cannot be transmitted until the supply pressure builds up to a certain level. The oscillator circuit using this is shown in fig. 6.1. Simple back-pressure sensors monitor valve position, pressures and temperatures. Other types are used for level and engine speed.

This system has been used to control a 300 kW land-based generator set. Encouraging results have been obtained, and the Admiralty is planning to build a fully engineered version which is intended for service at sea.

Another system, somewhat similar in principle, is under development by the U.S. Navy for the start-up of turbines to drive the main forced-draught blowers (*2*). So far the experimental work has been limited to the fluidic segment of the system, again using wall-attachment devices, the turbine being simulated on an analogue computer.

This takes the automatic control concept one step further than the Plessey system does, for the plan is to have pneumatic valves operated by the system in place of manual valves.

The U.S.S.R. is also active in this field, but the published paper (*3*) unfortunately gives very little detail. Moving-part diaphragm devices are used for the logic functions. The devices have reached a high state of development in the U.S.S.R.

These three applications do illustrate areas where fluidics can compete successfully with electronics. The shipboard environment, whilst not ideal, is not severe for electronic equipment. From the cost point of view the logic component of electronic systems is

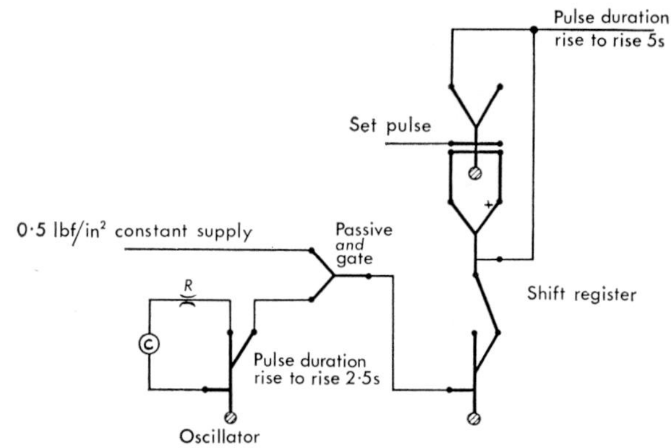

Figure 6.1 Oscillator and divide-by-two layout (*1*).

probably as cheap as, if not cheaper than, that of fluidics. Although these applications employ only a moderate amount of logic they do have a relatively large number of sensor inputs. It is here, the input interface, that fluidics scores, for the sensors employed are in the main cheap, reliable and tolerant of environment.

A slightly different diesel control system is being developed by Westinghouse Brake and Signal Co. (*4*, *5*), for the remote bridge control of the main propulsion engine. It is intended to be a replacement for their standard electropneumatic system. One of the advantages of fluidics in this case is the considerable reduction in size of the control panel. The standard system measures about 3 m by 3·6 m (10 ft by 12 ft) whereas the fluidic panel is about 0·45 m by 0·45 m (18 in by 18 in). At present the system, a laboratory prototype, only senses output-shaft speed and direction but others are under development for monitoring fault conditions. Sea trials of a second prototype are shortly to be carried out in a small cargo vessel.

6.2 Servo-control systems

Several servo-systems are at various stages of development in the U.S. Details of five projects have been published. These are concerned with ships' boilers, hydrofoils, torpedoes and turbines. Four are being developed by Bowles Engineering. General Electric is responsible for one of the turbine speed control systems.

The boiler control system (*6, 7*) is probably in the most advanced state of development. It is to replace the conventional pneumatic

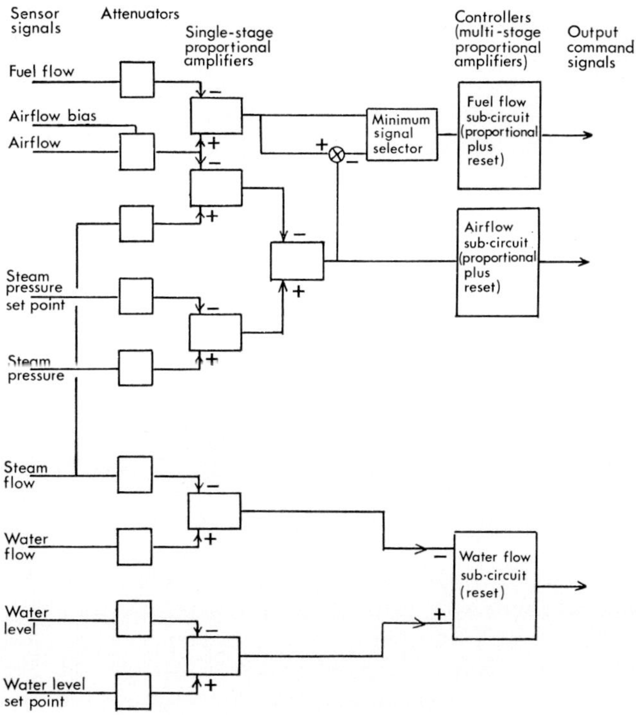

Figure 6.2 Boiler control system.

controllers in current use. In effect, two interconnected systems control both steam pressure and boiler water level by varying fuel, air and water feeds to the boiler (fig. 6.2).

Input signals, other than set points, are obtained from conventional sensors. Their outputs are attenuated to a level suitable for fluidics. The sensors, which are concerned with related parameters, are calibrated to give equal signals when those parameters are in their correct relationships. For example, when the mass flow of steam and water are the same their sensor signals are equal. Likewise, when fuel and air flows are in the proportion for complete combustion, their sensor signals are also equal. Thus, when the boiler runs in a steady state, the single-stage momentum-interaction proportional amplifiers operate in their null condition. This minimizes non-linear effects. The purpose of the minimum-signal selector is to prevent the excessive fuel flow which would otherwise follow a sudden increase in operating level because the fuel valves have a faster response than the blowers. The controller sub-circuits are multi-stage proportional amplifiers, capacitors being used to provide the reset (i.e. integral) action.

The elements and sub-circuits are cast in metal-filled epoxy resin using aluminium manifold plates. The various modules are rack-mounted and interconnected with copper tubing, a typical naval requirement. Tests with a naval boiler 'show that it performs as well or better than a conventional type of boiler control'. However, a report from the U.S. Naval Ship Engineering Center (*8*) shows that some problems were experienced with transmission-line delays and impedance matching between amplifiers. It recommended development of high-gain feedback amplifiers for use in a manner similar to that of electronic operational amplifiers. Consequent work has apparently reached fruition as such amplifiers are now commercially available.

The *torpedo course-control system*. Only brief details have been published of the sensor and associated circuitry (*9*). A vortex rate-sensor (cf. Section 1.8) generates a signal proportional to the torpedo's rate of turn. The two sensor outputs each feed a pressure-controlled oscillator with outputs which are fed into a difference counter. The counter output is a measure of the torpedo's angle of turn.

The *hydrofoil control system*. An ordinary hydrofoil tends to follow the contours of the water surface, and in rough seas this gives a poor ride and can even lead to instability. The purpose of the hydrofoil control system (*10*) is to keep the craft on a level course. The angle of attack of the foil is inferred from the differential pressure measured between two pitot tubes set symmetrically into the foil above and below the centre-line. The pressure signal is compared with a reference signal and the difference—which is proportional to angular error—amplified by a six-stage proportional amplifier circuit. Feedback round three of the stages is used for lead-lag compensation. The output stage driving the jack to vary the foil angle is also a proportional amplifier. This is unusual as a pneumatic-fluidic or hydraulic-fluidic step-up valve is normally used to conserve power. The continuous power drain of a fluidic output is acceptable only for low-power drives or for intermittent use.

The two *turbine-speed control-systems* are interesting in that they represent slightly different approaches to the same problem, the control of a 500 kW 12 000 rev/min turbine generator set to an accuracy of $\pm 0{\cdot}5\%$. Although in their initial development the systems used air, in service they will operate on steam at 480 $\pm$ 28°C (900 $\pm$ 50°F). This permits location in the steam turbine casing. Reliability is further increased because the system will not depend on the operation of other auxiliaries for its power supply.

In the Bowles system (*11*) a disc with four holes in it, mounted on

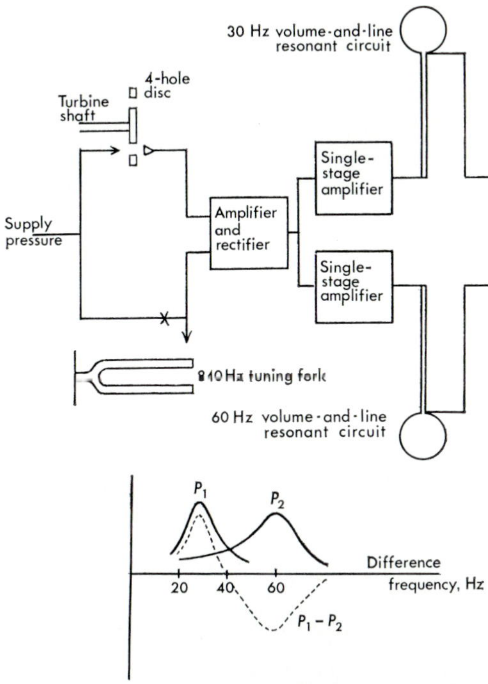

Figure 6.3 Bowles turbine-speed sensor.

the shaft, is used to generate a signal at a nominal frequency of 800 Hz. Another signal, at a frequency of 840 Hz, is obtained from a tuning-fork oscillator. The two signals are fed into the control ports of a proportional amplifier, the output of which is rectified to give the difference frequency. The deviation of the output from its nominal value of 40 Hz is a measure of the speed error. The signal is fed to two resonant circuits forming the discriminator (fig. 6.3). The circuits use the inductive effect of the inertia of the fluid in a line with the capacitance of a volume. The output of each resonant circuit is amplified and rectified. Subtraction of one rectified signal from another gives a differential pressure which is a function of the speed error. This signal is amplified and used to drive an actuator as in the hydrofoil circuit.

The General Electric approach (*12*) differs from the Bowles system mainly in the form of the sensor and discriminator. A wobble plate with back-pressure sensors produces a sinusoidal signal at

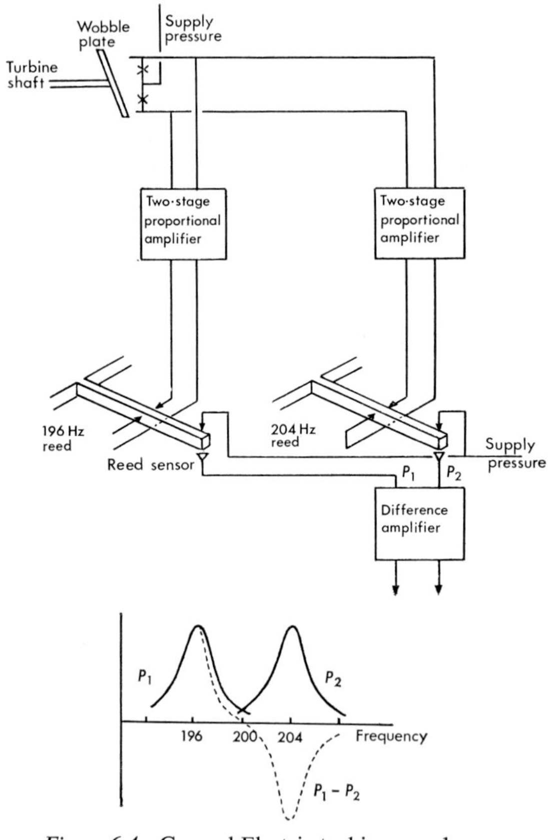

Figure 6.4 General Electric turbine-speed sensor.

shaft speed. This is amplified and used to drive two vibrating reeds (fig. 6.4), one with a resonant frequency of 196 Hz and the other 204 Hz. At nominal speed, 200 Hz, their amplitudes of vibration are equal. If speed deviates from nominal the amplitude of one vibration, and hence the mean pressure of its sensor, is greater than the other. Subtraction of one signal from the other gives a differential pressure proportional to speed error.

An important factor affecting the design of the systems is the variation in temperature of the steam. In the General Electric system this causes a small error owing to the change in resonant frequency of the reeds. The tuning fork in the Bowles system is similarly affected but it has an additional error—the variation with temperature of the resonant frequency of the two tuned circuits. The selfsame mechanism provides the basis of a fluidic temperature sensor. The General Electric system has been delivered to the U.S. Navy Engineering Center.

6.3 Large water-operated device

So far interest in water-operated devices has been limited to their use as thrusters. Bow thrusters are installed in ships for manœuvring in confined spaces. The conventional screw bow thruster is not easy to maintain and is difficult to install in existing ships. A jet thruster, using the reaction force of a water jet to provide the thrust, is far simpler. Although its efficiency is low that can usually be tolerated as bow thrusters are used only intermittently. Also pumping equipment installed for other purposes can sometimes be used to power the jet.

Bowles has developed a simple bow thruster for use in barges (*13*). It is based on a wall-attachment amplifier (fig. 6.5). If both control ports are uncovered and inducing air the water stream does not attach itself to either wall and divides itself evenly between the two output legs, giving zero net output thrust. Covering one of the control ports causes the stream to attach itself to the wall and directly downstream of that port pass into the appropriate output channel, giving full thrust in one direction. Only the overall dimensions of the device have been published, but a photograph has shown that the outlet dimensions are of the order of 18 cm by 54 cm (7 in by 21 in). The quoted thrust is 7600 N (1700 lbf) which would be obtained with a flow rate of about 830 l/s (11 000 gal/min) at pressures of about 0·04 N/mm² (6 lbf/in²).

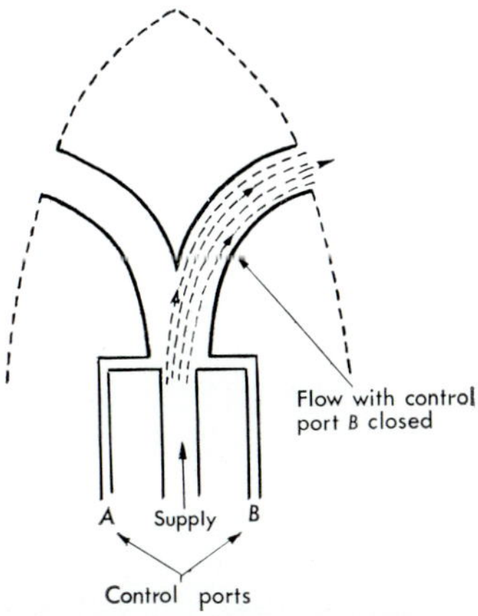

Figure 6.5 Wall-attachment bow thruster.

The disadvantage of the wall-attachment thruster is that only zero or full thrust can be obtained. This gives very coarse control, and furthermore ships' body-bending vibration modes could be excited by the 'bang-bang' action. A thruster based on a proportional instead of a wall-attachment amplifier does not have these problems.

Bowles has built a proportional thruster for the U.S. Navy (*14*) and the U.S. Navy has also built one itself (*15*). However, the proportional thruster does not have the same simplicity of control as the wall-attachment device because relatively large control flows have to be supplied.

Bowles drives the thruster from another proportional amplifier, which is also powered by the water supply to the thruster to reduce the control flows (fig. 6.6). It would be possible to use several stages of amplification, all powered from the water supply, and to control the first stage with air from a fluidic control system. Another device, the axisymmetric focused-jet amplifier, has been investigated as a thruster (*16*, *17*). A single device of this kind gives only a unidirectional thrust.

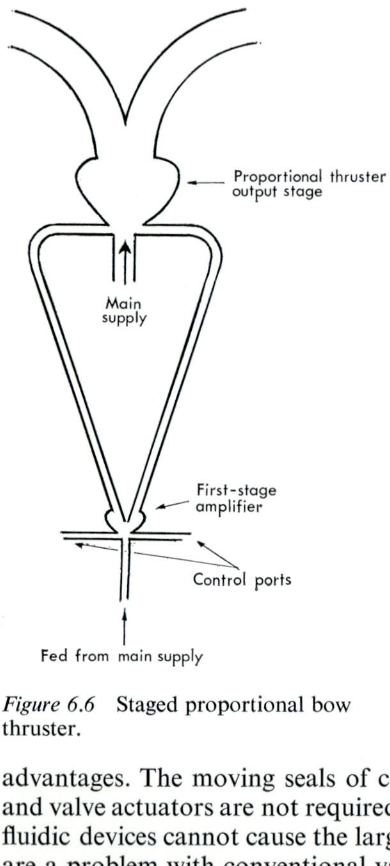

Figure 6.6 Staged proportional bow thruster.

With regard to the control of jet thrusters fluidics is competing with moving-part valves. The main disadvantage of fluidics is that the power consumption is about 20% to 30% higher than with moving-part valves. Against that there are several advantages. The moving seals of conventional valves are eliminated and valve actuators are not required. Furthermore rapid switching of fluidic devices cannot cause the large water-hammer pressures which are a problem with conventional valves.

6.4 Problem areas

The problems in marine applications are similar to those encountered in other areas. However, they do reflect particularly the need in marine equipment for very high reliability in a not very severe environment, rather than a demand for low cost or the ability to operate in severe environments. The two most important problems in this respect appear to be the reject rate after manufacture and contamination of the air supply.

A reject rate as high as 5% after delivery of equipment has been quoted for individual devices (*8*). Unless manufacturing techniques improve, the implications are that the reject rate for integrated circuits will be considerably higher. Several techniques for overcoming this are already in use, among them the manufacture of circuits separately from the elements which are inserted into the circuits, and the bonding of elements or their mounting with O-ring seals on the manifold. It remains to be proved that, by these techniques, circuits can be produced which have the required high reliability.

Contamination is the more serious problem. It was first thought that the requirements for a fluidic air supply would not be too stringent as any contaminant that passes through a normal air filter would also pass through the devices. This is true, within limits, of dry particulate matter. But vapour, and particles of oil less than 1 μm in diameter, which are present in most air supplies, will pass through a filter. Some of these passengers are deposited in the interaction region of amplifiers, building up sometimes a lacquer and at other times a greasy layer of dust and dirt, which eventually cause failure. Difficulties were encountered with the boiler control system after seven months' operation with an air supply more oil-free than the ambient air (*8*).

It is encouraging that much less trouble has been experienced in industrial applications, where 'instrument-quality' air is used (*18*). Oil filters designed with this problem in mind have recently been marketed. With improved air supplies life will be extended considerably.

But it is unlikely that fluidic devices will last the life of a ship without any attention. Routine cleaning will probably be necessary, perhaps every year or so. Chemical flushing is likely to be the most suitable. The choice of manufacturing method will be affected because some of the plastics in current use may not have the requisite chemical stability. Integrated circuits which cannot be dismantled are much more difficult to clean than individual elements as it is almost impossible to ensure that the cleaning fluid gets into every passage in the block.

6.5 Future prospects

The future of fluidics in the marine world depends to a large extent on whether the present developments prove themselves in use as standard equipment. So far indications are favourable, and it is to be expected that fluidic systems will gradually come into service over the next five years.

In the near future defence applications will continue to predominate. As with any new technique there is speculation on other ways in which it can be used. Two feasibility studies give examples of this. One is concerned with the control of pontoons used in salvage work (*19*) and the other with the hovering control of submarines (*15*). In the latter case it has been suggested that the control as well as the thruster fluidics operate with water as the working fluid. There may be difficulties as the response times are longer with water than with air.

Commercial use will probably follow the successful development of defence systems. Fluidics is truly a case where ‘spin-off’ applies, since development costs are high and hardware costs are low. With the present emphasis on reduced manning and more intensive use of ships there are many potential applications.

7

AEROSPACE APPLICATIONS

G. C. GANT
British Aircraft Corporation Ltd

Because of the connexion between aerospace research and national security much of the work that is being done is classified. Consequently performance data in some areas are rather scarce or out of date. Nevertheless, quite a few very interesting applications have been reported, and the following paragraphs describe a selection of these to give what is believed to be a fair representation of the scope of fluidics in aerospace.

Fluidic technique is very promising for aerospace control. Fluidic

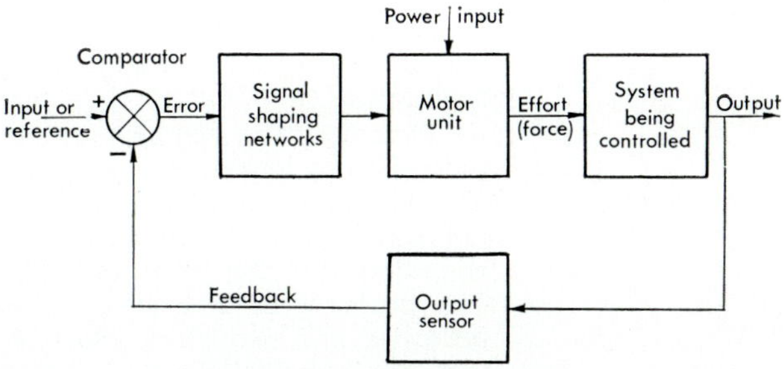

Figure 7.1 General control loop.

devices can in principle withstand a wide range of hostile environmental conditions. Examples are high temperature, nuclear, cosmic and electromagnetic radiation, high acceleration and vibration, and corrosive atmospheres. Most of these are very upsetting to electromechanical systems. Furthermore, fluidics offers advantages in reliability. Moving parts are not used, and there is no effect on radio communication. It should not be forgotten, though, that fluidics cannot measure up to electronics for accuracy and speed of response. In my view fluidics should not be considered as a replacement for other techniques but as a complement to them, with its own particular spheres of application.

Glossary of control terms

CONTROL LOOP	A general name for a control system in which the output is compared with the input and the error is used to correct the output. See fig. 7.1
OUTPUT	A controlled variable such as the position of a mass or the speed of an engine
INPUT	The signal fed into a control system to change the output
ERROR SIGNAL	The difference between the input and output
COMPARATOR	The circuit or mechanism which compares input and output to produce an error signal
DAMPER	A device for reducing an over-large rate (such as an excessive velocity)
INTEGRATOR	A device which adds its inputs as a linear function of time (mathematical integration)
GAIN	The dimensionless ratio of the magnitudes of the output and the input of a device
RC LAG NETWORK	By electrical analogy, a circuit that passes low-frequency signals while attenuating high-frequency signals
TIME CONSTANT	A time which is characteristic of the speed of response of a control device or system and which arises in the dynamic equations. In the case of a simple lag circuit the time constant is the product of the resistance and capacitance (*RC*) and is the time required for the output to reach 63% of the magnitude of a step input
GAIN SATURATION	The state of an amplifier in which the output remains constant at a maximum level regardless of further increases in input
STABILITY	The property of a system whereby the deflective and oscillatory effect of a disturbance is made to die out in a finite time
PHASE ADVANCE (LEAD-LAG) NETWORK	A network that passes high-frequency signals while attenuating low-frequency signals
SHAPING (OR COMPENSATION)	Modification of a signal, as in phase-advancing, usually to improve system stability, accuracy or response
FEEDBACK	Signalling a system output back for comparison with the input
MODULATOR	A device which varies some property, such as the pulse width, of a waveform in proportion to a signal
BANDWIDTH	The frequency range over which the magnitude of the output of a system is not less than some set figure, commonly 70% of the input magnitude

TRANSIENT RESPONSE	The time response of a system, seen as the output following a standard input
HIGH-PASS FILTER	A network that passes high-frequency signals while attenuating low-frequency signals. It is the same as a lead-lag phase-advance network but with no steady-state reference

7.1 Missile control

Various missile systems have been mechanized. The main differences lie in the methods of applying the controlling forces (the effort) to the missile. Three means have been extensively tried: thrusters (reaction jets), thrust vector control, and jet flaps. Applications of these methods to directional and roll control are described in the following sections.

There is an important feature of missile systems which sets them apart from most others. They usually need, for convenience and power, to operate on 'warm' or 'hot' gas. In this context 'warm' gas is usually at 1000 to 1200°C while 'hot' gas is usually at 2500 to 3000°C. 'Warm' gas is generally produced by solid-propellant auxiliary gas generators while 'hot' gas is tapped from the main rocket motor.

A wide range of propellants may be used to produce high-temperature gas, and the product may contain varying amounts of soot and solid particles. So two major problems arise in the design of fluidic systems for missile control—the high temperature and contamination levels of the gas supply. If the high-temperature gas can be used, perhaps mixed with ram air, missiles are natural test beds for fluidic devices. The main saving grace of a missile is that it is usually only a one-shot device of very short life (a few seconds). This eases the problems of operation on high-temperature gas because system hardware seldom reaches the gas temperature.

7.1.1 *Roll control using thrusters*

A roll control system using thrusters has been mechanized and flight-tested by the U.S. Army (*1*). A roll control system was chosen as simplest for the first attempts but a similar system could readily be used for guidance of a naturally stable missile.

The mode of operation was on/off. An on/off system is very non-linear. In the presence of an error signal above a certain threshold level, the control force is switched hard on in one direction and switched off again when the error is reduced. In the absence of an error signal the control force switches on and off continually in opposite directions in an oscillation called a limit cycle, the limit cycle frequency and amplitude being determined by the system parameters. The frequency of the system tested was about 3 Hz.

The block diagram for the on/off system mechanized and flown is shown in fig. 7.2. It is a roll rate damper system with an integrator to minimize roll attitude drift.

The form of the fluidic circuit inferred from the published

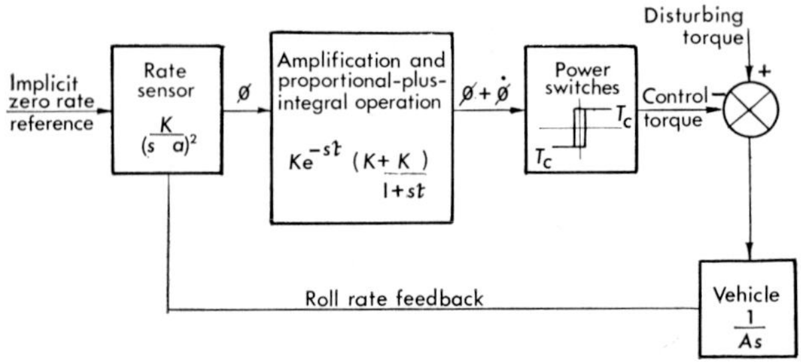

Figure 7.2 Roll-rate damper system.

schematic diagram (*1*) is shown in fig. 7.3. For simplicity only single-stage fluidic amplifiers are drawn, but in fact multi-stage amplifiers are used to give sufficient gain.

The rate sensor signals the direction in which, and the rate at which, the vehicle is being rolled by the disturbing torque. The signal from

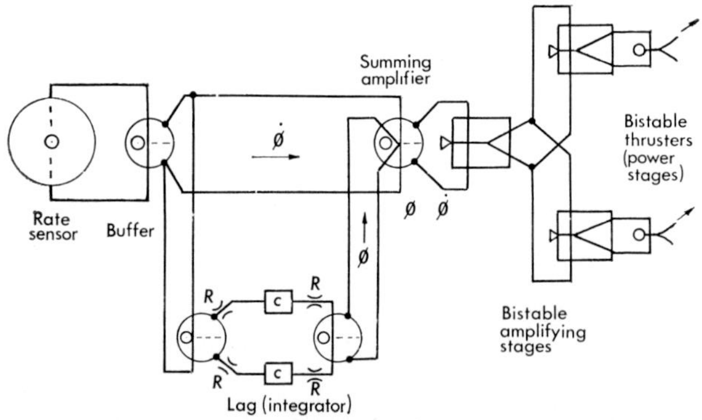

Figure 7.3 On/off roll-control circuit.

the rate sensor buffer amplifier is integrated in an *RC* type of pneumatic lag circuit with a 10 s time constant to give a position signal ϕ. (The capacities *C* are tanks of about 1600 cm³ volume with restrictors *R* of about 0·3 mm diameter.) The integrated signal (position ϕ) is added to the rate signal ($\dot{\phi}$) and the sum is fed to a bistable stage which switches when the signal from the summing amplifier exceeds the switch threshold. The resulting on/off signal is then amplified to a power level sufficient to switch the thrust jets, the final stages being supersonic elements.*

* A supersonic element is one in which the driving pressure ratio is such that the gas flow velocity through the element exceeds the local speed of sound.

The thrusters are arranged so that the reaction force due to the gas efflux from the outlets acts tangentially to the missile body circumference, giving rise to a torque opposing rotation, thus damping the undesired motion.

A passive lag was chosen rather than a pure integrator for this system. The advantage is simplicity, since fluidic integrators require precise adjustment and are very sensitive to temperature and supply pressure changes.

The system is supplied from stored nitrogen at 24 MN/m^2 pressure. For the signal circuits, amplifying circuits and power stages respectively the pressure is regulated down to 24 kN/m^2, 83 kN/m^2 and 1·83 MN/m^2. The temperature drops on expansion of the gas from storage to signal pressure, and to overcome adverse effects on system operation a piston type of temperature buffer is included.

As an alternative to the on/off system a variable-mark/space-ratio mode of operation could be used. A variable-mark/space-ratio system is one in which the control force is deliberately switched on and off at a predetermined frequency and the difference between on and off times (the pulse width) is varied in a continuous manner. Control is therefore linear, with some dither up to the saturation limits of error, when the control force remains fully on. This is gain saturation, and the rate of response of the system thus has a definite limit. The method of control is described below, in connexion with some of the systems that use it.

7.1.2 *Steering using jet flaps*

A true missile steering system, as distinct from a guidance system, is under development by the British Aircraft Corporation at Stevenage. Jet flaps are being used in this work to steer a hypothetical anti-tank weapon (*2*).

Jet flaps are sheets of gas issuing from the trailing edges of wings, with flap slots formed by the outlets of high-aspect-ratio fluidic switches built into the wings.

The operation of jet flaps in subsonic flow is to promote circulation round the wing, in the same manner as does a mechanical flap or aileron, and hence cause lift. The magnitude of the lift force measured on a low-aspect-ratio wing has been up to nineteen times the reaction thrust of the flap. In this way adequate steering forces can be applied to the missile with the expenditure of quite a small quantity of gas—about 8% of main rocket thrust.

The system under test is a variable-mark/space-ratio system controlling missile heading in response to signals from the ground. It differs from previously described systems in a number of respects apart from the use of jet flaps. For example, the gas supply is a mixture of main rocket-motor gas and ram air. Also it is a combination of electronic and fluidic elements. And finally, it is a servomechanism rather than a regulator, that is it responds to a varying

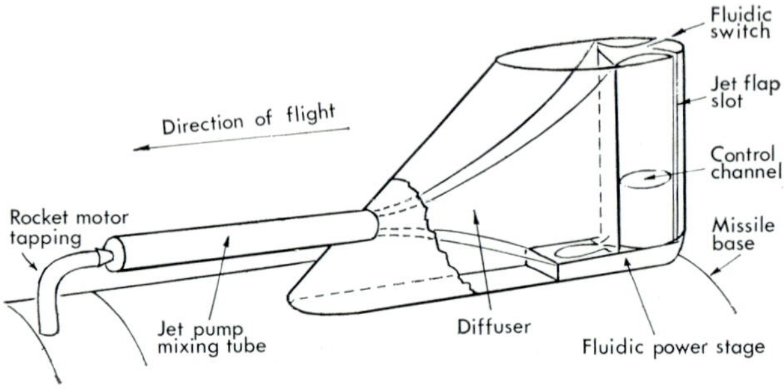

Figure 7.4 Wing with jet flap switch and jet pump.

command and does not merely maintain a situation against external disturbances.

The gas supply being developed for this system uses jet pumps, one in each wing. These mix hot high-pressure rocket-motor gas with cold ram air in roughly equal proportions to produce a large-volume flow of quite cool (400°C) low-pressure (55 kN/m^2) gas which

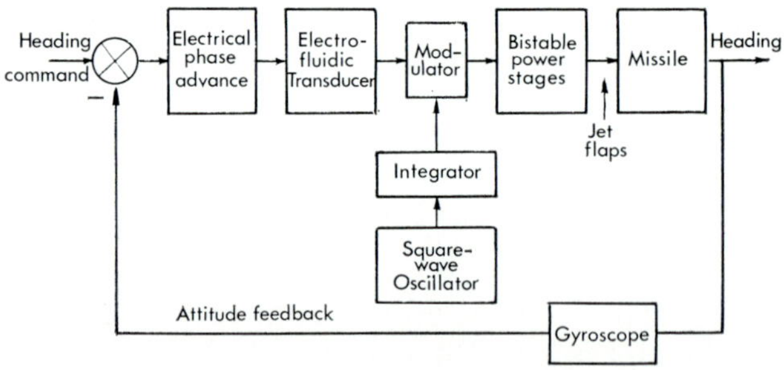

Figure 7.5 Missile steering system with jet flaps.

is suitable for fluidic circuitry. The jet pumps are manifolded together for balance and the gas supply for the fluidic circuits is taken from the manifold. For simplicity and reliability only bistable fluidic elements are used in the system.

The arrangement of wing, jet flap switch and jet pump is shown in fig. 7.4.

The block diagram for the system is shown in fig. 7.5. The missile has to be highly manœuvrable so it has very little natural stability. In order to stabilize it to the right degree a phase-advance shaping block is included. This is electronic.

Some electronic gear is used in the system because command signals are received by the missile in electrical form and then converted into fluidic form for operation of the jet flaps.

Although work is proceeding to evaluate fluidic phase-advance circuits and systems with rate feedback from a vortex rate sensor, the principal activity is on a compound electronic–fluidic system with the electronics inside the control loop. This is felt to be the system most promising of early success since it can include existing and proven position gyroscopes with electrical pick-off and it can ensure accurate phase-advance of the signal.

One axis of control circuitry is shown in fig. 7.6, in which a freely rolling missile is assumed. It is also assumed that roll is resolved on

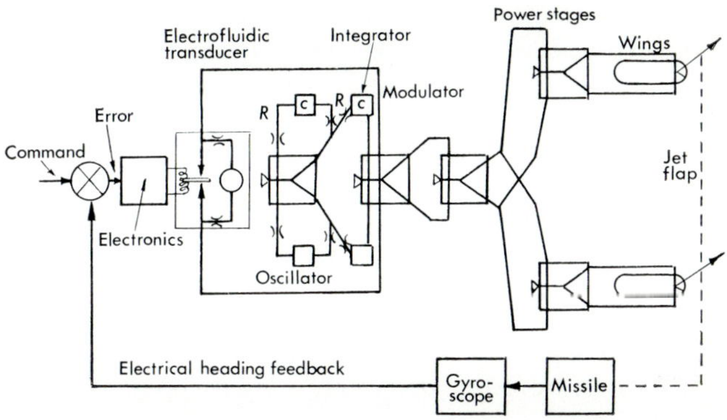

Figure 7.6 Jet-flap control circuit.

the ground, which means that the command guidance signals are fed to the right channel at the right time to give control in azimuth and elevation relative to the operator. Roll control could easily be arranged by having a separate channel for each wing, though much more hardware would then have to be accommodated. As can be seen in fig. 7.6, the electrical heading command signal is compared with an electrical heading feedback signal from the gyroscope. The error signal is phase-advanced, amplified and fed to the transducer. The signal from the transducer is fed into one pair of inputs of a four-input bistable fluidic element and a triangular wave is fed into the other pair of inputs. The output from the four-input modulator element is a rectangular wave with pulse-width proportional to the shaped error signal. This is amplified in two power stages to drive the jet flap switches.

The transducer used in the prototype circuit is of the flapper-nozzle type. A flapper moves between two back-pressure nozzles which are fed through fixed restrictors, and the back-pressure between each nozzle and its restrictor varies with flapper position. From the two nozzles a differential pressure, proportional to electric current input, is obtained for use as the signal. A combination of two nozzles in a differential arrangement of this sort is superior to a single nozzle because differential linearity is better than one-sided linearity and the sensitivity is doubled. The device is analogous to the electrical engineer's Wheatstone bridge.

The oscillator uses positive *RC* feedback round a bistable element, as shown in fig. 7.6, and the output is a square wave adjusted for unity mark/space ratio. The square wave is passed into a lag network which has a time constant such that the network acts as an integrator at the oscillator frequency (50–60 Hz). This produces the triangular wave. One oscillator is used for both channels to define the cross-coupling.

The whole fluidic system works on a common supply pressure. Power is amplified by increasing the size of successive stages. The power stages are not vented, being matched together, but the smaller elements in the signal circuitry are of standard vented form to facilitate matching.

A closed-loop bandwidth of 2 Hz is readily achieved with the fluidic system, and the radius of turn and transient responses are as good as those of conventional weapons. In some areas the performance promises an improvement over existing systems with much lower levels of structually transmitted noise. The lower noise levels are important for the better operation of gyroscopes.

7.1.3 *Guidance under thrust-vector control*

Thrust-vector control or t.v.c. is exercised by varying the line or vector of thrust of the main rocket-motor relative to the vehicle

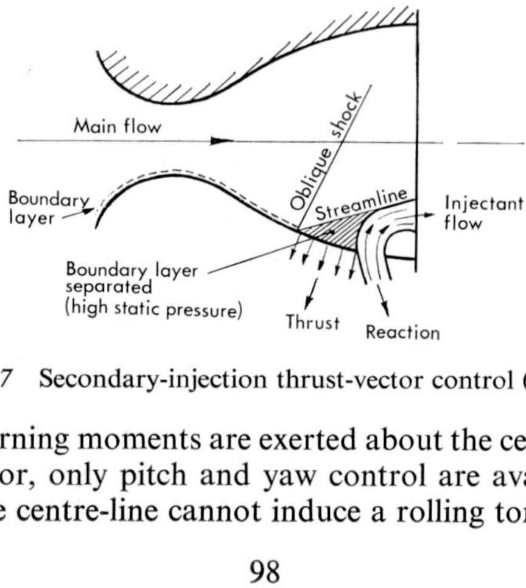

Figure 7.7 Secondary-injection thrust-vector control (s.i.t.v.c.).

axis, so that turning moments are exerted about the centre of gravity. With one motor, only pitch and yaw control are available because a thrust on the centre-line cannot induce a rolling torque. If several motors are clustered round the centre-line, as in most space rockets, then of course roll-control is possible.

T.v.c. methods vary widely, each having its own applications and limitations. The kind which lends itself best to fluidic technique is known as secondary-injection t.v.c. (or s.i.t.v.c.). A liquid or gas is injected into the main-rocket gas-stream through ports in the divergent part of the nozzle. This causes boundary layer separation and the formation of an oblique shock wave. The result is a side force with two complementary components, as shown in fig. 7.7. They are a reaction force from the inflowing jet and a fluid thrust on the nozzle wall. The side force exerted is greater than that which would result from exhausting the injectant through a thruster.

A magnification factor has been defined by Rivard (*3*) as—

$$K = \frac{F_s}{\dot{M}_s} \times \frac{\dot{M}_a}{F_a}$$

where F_s = side force, F_a = axial thrust, $\dot{M}_s$ = injectant mass flow rate and $\dot{M}_a$ = rocket motor mass flow rate. This factor can vary from about 0·3 to over 2·0, depending on the nature of the injectant. At the low end of the scale are non-reactive cold liquids and at the top of the scale are hot gases. For greatest magnification the injectant gas should have a low molecular weight and low specific heat. The efficiency is of course also affected by the position and direction of the injected flow.

An interesting method of controlling the injectant flow has been reported by Bendix Research Laboratories (*3*). Fluidic vortex valves are used to control warm- and hot-gas injectants. Here the main advantage of vortex or any other types of fluidic valve is that they are insensitive to the thermal transients which seriously upset moving part valves. The vortex valve is really an amplifier used to throttle injectant flow.

An s.i.t.v.c. system tested with warm-gas injectant is shown in fig. 7.8. It operates in a push-pull mode. That is, when no steering force is required the effects of the flow of injectant gas from the two valves are balanced, but not so if a demand arises for change in direction.

For example, should a reduction in the output of valve 1 and an increase in the output of valve 2 be required, an electrical signal closes the flapper nozzle in proportion and increases the control flow to valve 1. This reduces the output from valve 1, which increases the impedance to the supply flow and raises the supply pressure. The bias pressure to valve 2, which would otherwise depend on supply pressure, is kept constant by a relief valve. Owing to the increased supply pressure the supply flow through valve 2 increases. The gas flow taken from the gas generator is roughly constant.

Static tests with 1100°C gas on this system have demonstrated a

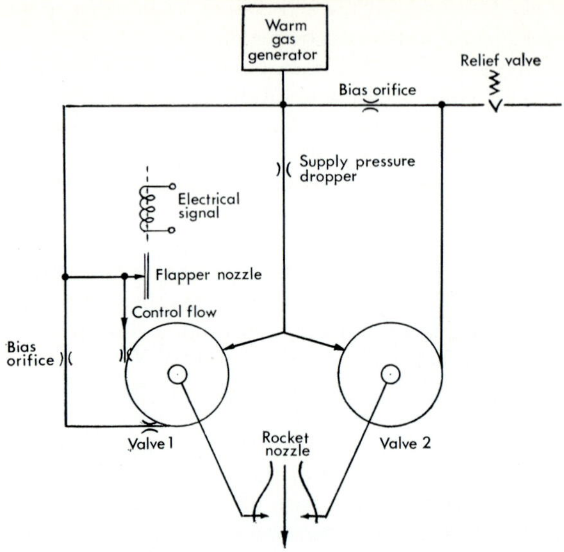

Figure 7.8 Warm-gas s.i.t.v.c.

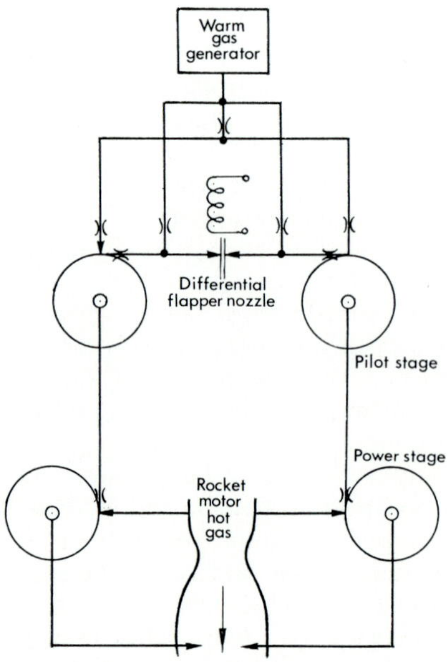

Figure 7.9 Hot-gas s.i.t.v.c.

flow turn-down ratio in excess of 4 : 1 with stable operation. Sinusoidal frequency response showed a bandwidth (−4dB, 28° lag) of 30 Hz. The major limitations on response were in the manifold volumes. Similar systems have shown a bandwidth in excess of 85 Hz. The transient response of output pressure to a step input to valve 1 was about 5 ms.

It has been shown that s.i.t.v.c. can be more effective using higher-temperature gas—for convenience directly bled from the rocket motor. Some work has been reported (*3*) on such a system with two stages; a pilot stage working on warm gas and a power stage working on hot gas, as shown in fig. 7.9.

The mode of operation is again push-pull to bleed the warm gas generator at a constant rate. An electrical signal operates the flapper-nozzle and unbalances the pilot-stage output flow. This in turn unbalances the power-stage output flows, shifting the net thrust-vector.

The design of power stages to operate on the hot rocket gas requires a lot of care. The gas has in suspension aluminium oxide droplets which are extremely erosive and which 'plate out' on cold surfaces, tending to block valve passages at ignition. The high-velocity flow regions of the valves are made from silver-impregnated tungsten to combat erosion, and the low-velocity regions are lined with carbon-phenolic material, which is a heat insulator, to minimize plating-out. The general structural material is stainless steel.

In any application of such a system the final-stage valves would be mounted on the rocket nozzle, which might be a submerged type, to avoid need for manifolded supplies.

The injectant could, of course, be controlled with momentum-interaction amplifiers. Work carried out at B.A.C. Bristol on this topic has been reported by Platt (*4*). The systems using vortex valves are chosen for description here because of their novelty.

7.2 Aircraft controls

Numbers of aircraft fluidic control systems have been flight-tested, commencing with a pitch-rate damper system on a single-engined light aircraft in 1963. Since then systems of increasing complexity have been tested, including a three-axis damper for a helicopter and a complete three-axis autopilot for a twin-engined aircraft. Space permits the description of only one aircraft fluidic control system.

The three-axis autopilot which was sponsored by N.A.S.A. and designed by Honeywell Inc. (*5*) emerged from a development programme to provide experience of fluidic flight-control systems and to lay the foundations for the future development of low-cost automatic flight control for light aircraft. The test-bed aircraft chosen was an Aero Commander 680 FP, and the system was designed to damp yaw, level the wings, command heading, hold heading and hold altitude.

The systems are arranged to operate in parallel with the pilot's manual controls and there is a built-in mechanical override for safety. The aircraft is controlled through conventional control surfaces (ailerons, elevators and rudder) which are operated by piston actuators driven by final-stage fluidic power-amplifiers. Servo-output movements are reflected in movements of the control wheel and column and rudder pedals.

Aircraft control tends to be less straightforward than missile control because an aircraft must bank to turn. The control system developed consists of two multi-loop systems, one providing lateral control (yaw and roll) and one longitudinal control (pitch and altitude).

7.2.1 *Lateral axis control*

The lateral axis block diagram is shown in fig. 7.10. Yaw rate and heading are sensed and heading command, heading hold, yaw damping and wing levelling (roll control) are provided. The wings are

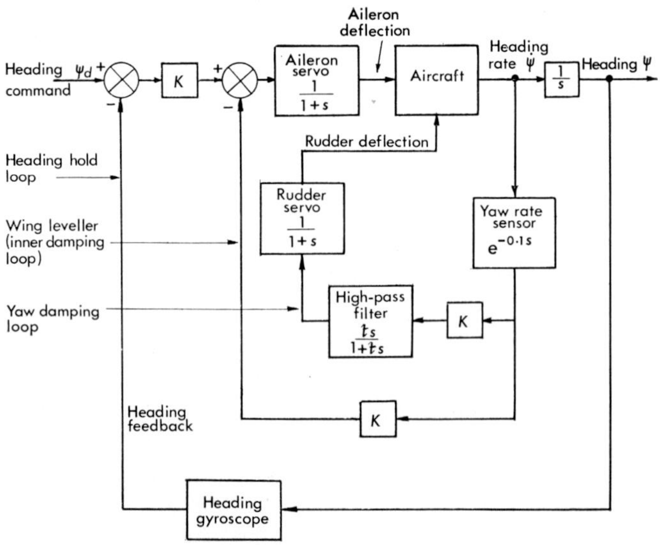

Figure 7.10 Lateral-axis control system.

levelled by feeding yaw rate to the ailerons. Since roll angle is directly proportional to yaw rate (for constant speed and altitude), roll attitude is maintained without roll attitude reference.

The fluidic circuit corresponding to the block diagram is shown in fig. 7.11. Yaw rate is measured by a fluidic vortex rate sensor and summed at the aileron servo-mechanism to form the wing-leveller loop. Aircraft heading is sensed with a directional (inertial) gyroscope with pneumatic pick-off and a signal is fed back in the heading-holding loop. The gyroscope has a heading selection feature so that heading commands can be made by the pilot.

The yaw rate signal is also shaped by a 2 s high-pass filter and fed to the rudder servo-mechanism forming the yaw-damping loop. The high-pass filter blocks steady-state signals and so allows the aircraft to make constant-rate heading changes unopposed by the rudder.

The control forces are again pulse-width-modulated. Proportional

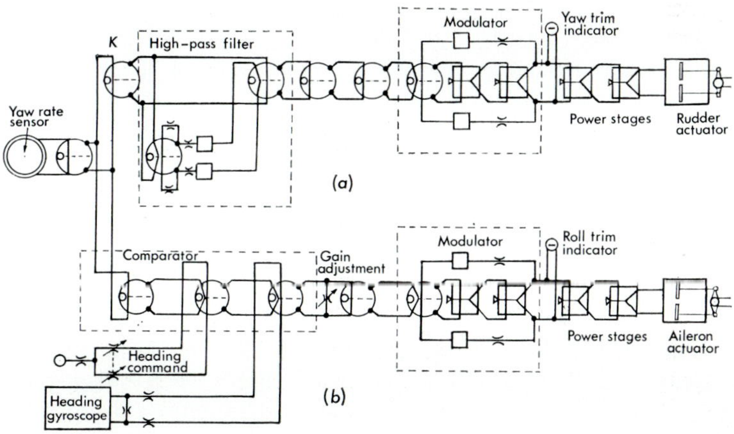

Figure 7.11 Lateral-axis control circuit. (a) Rudder-servo yaw-damping loop. (b) Aileron-servo wing-leveller and heading hold.

all-fluidic circuits are used for signal manipulation. Another type of modulator is used in this circuit. It works by biasing an oscillator with the error signal. A separate oscillator is used in each control channel.

Each oscillator/modulator consists of three stages. The first stage is a summing proportional amplifier and the second and third stages are bistable elements. Positive *RC* feedback is applied round all three stages, and the values of *R* and *C* selected to give a carrier frequency of about 25 Hz.

The 2 s high-pass filter action is obtained by passing the signal through a feedforward lag network and adding it to the original signal. At low frequencies the signal is cancelled out by the feed-forward signal but at high frequencies the signal passes through unaffected. The restrictors and capacities of the lag circuit are approximately 0·18 mm in diameter and 260 cm^3 in volume. Negative feedback is employed round the feedforward amplifier to improve linearity.

The modulated signal in each channel is amplified and fed into an opposed-piston actuator which moves in proportion to the pulse width of the incoming signal. Some carrier-frequency dither at the actuator would be useful to keep the actuator 'alive', and so avoid Coulomb (or static) friction. The actuator rotates a cable drum and the cable operates the control surfaces.

7.2.2 *Longitudinal axis control*

The longitudinal axis block diagram is shown in fig. 7.12. Here the pitch rate is measured with a fluidic vortex rate-sensor and signals

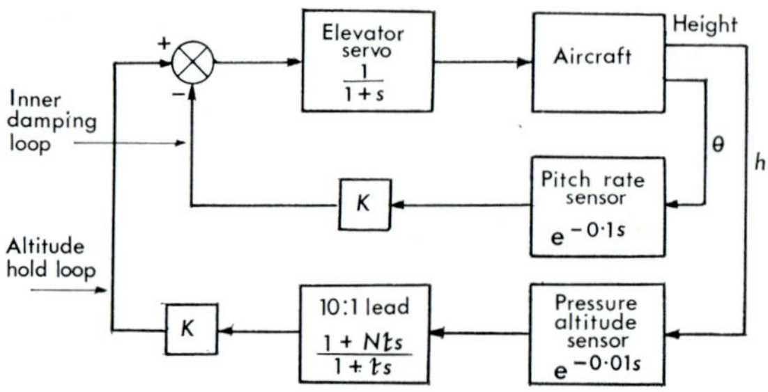

Figure 7.12 Longitudinal-axis control system.

are summed at the elevator servo-mechanism to form an inner damping loop. Pressure altitude is sensed, shaped with a 10-to-1 lead network and added to the pitch rate signal to complete the altitude-

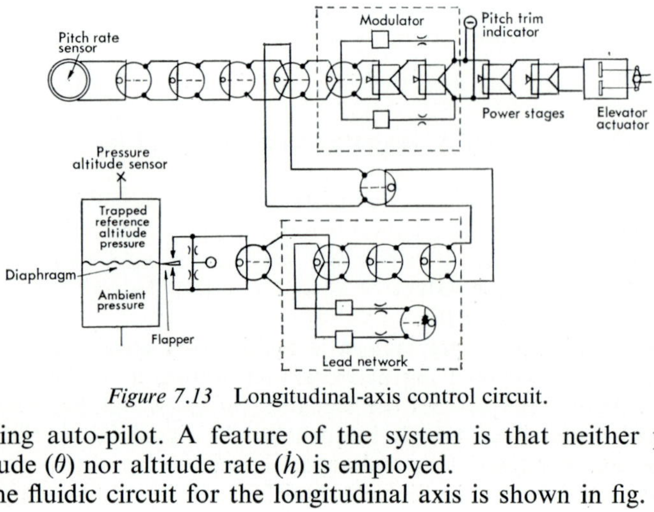

Figure 7.13 Longitudinal-axis control circuit.

holding auto-pilot. A feature of the system is that neither pitch attitude (θ) nor altitude rate ($\dot{h}$) is employed.

The fluidic circuit for the longitudinal axis is shown in fig. 7.13.

The interesting features of this circuit are the lead (phase advance) network and the altitude sensor.

The lead circuit consists of *RC* lag negative feedback round an amplifier, the feedback restrictors and capacitors being approximately 0·18 mm in diameter and 285 cm^3 in volume. Close agreement with the theoretical frequency response has been obtained.

The altitude sensor has a diaphragm on one side of which air is trapped at the pressure corresponding to the required altitude and the other side of which is open to atmosphere. A flapper is attached to the diaphragm. As the altitude changes the pressure difference across the diaphragm changes and shifts the flapper between two back-pressure nozzles, producing a differential pressure signal.

The sum of the pitch-rate and altitude signals is modulated and amplified to operate a cable drum actuator as described for the lateral axis.

7.2.3 *General features*

The proportional amplifiers in the system are of the vented beam-deflector (momentum interaction) type, operating on 10 kN/m^2 and 28 kN/m^2 air. The amplifiers are chemically etched in copper beryllium with 0·25 by 0·125 mm power nozzles.

The bistable elements are aluminium-filled epoxy cast devices with nozzle sizes varying from 0·5 by 0·38 mm to 2·5 by 1·0 mm.

Air for the system is provided by two engine-driven pumps. It is filtered and regulated to 120 kN/m^2 and then passed to the system, where it is further regulated to the lower pressure level for the signal circuits.

The autopilot, generally as described here, has been built and flight-tested and its performance shown to be as good as that of conventional autopilots up to 2000 m altitude. Above this altitude the performance deteriorated, mainly owing to inadequate gain.

7.3 Jet engine controls

Considerable effort and success have been reported with jet engine controls.

Two distinct types of system (conventional or fluidic) are used on engines: automatic start/stop sequencing controls and closed-loop controls for speed regulation etc. As hitherto in this chapter, only closed-loop systems are covered in the following account.

Much of the work in this field has been done by the General Electric Co. (*6*, *7*) and Honeywell Inc., firms which have successfully built and tested fluidic systems for controlling engine speed, compressor pressure ratio and turbine inlet temperature and for limiting speed and acceleration. However, work has also been reported (*8*, *9*) by Dowty Fuel Systems and the Plessey Co. on, respectively, pressure-ratio sensing and inlet-guide-vane control. Some interesting examples have been chosen from among these cases for more detailed description below.

In a pressure-ratio loop, a speed loop and a temperature loop studied by General Electric it was decided to retain conventional

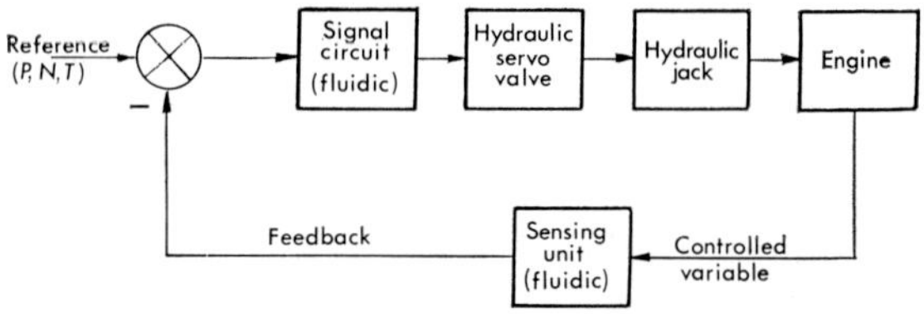

Figure 7.14 Jet-engine regulator.

hydraulic servo-valves and jacks for actuation and to use fluidics for sensing and signal circuits. A general block diagram for the three systems, which are all essentially regulators, is shown in fig. 7.14. The original purpose of the work was to compare fluidic with other control media. Fluidic systems showed up very favourably in terms of weight, size and failure rate, and this prompted further work on fluidic controls.

7.3.1 *Pressure-ratio loop*

The requirement here was to control the compressor pressure ratio as a function of the throttle position and the compressor inlet temperature. The pressure ratio is controlled by varying an airflow area in the engine.

The part of the system that it was decided to mechanize by pneumatic means was the pressure-ratio error-sensor and the amplifier. Following the impulse of most fluidic system designers, serious consideration was given to methods of achieving the required result using only static computational circuits. However, it became apparent that, owing to the complexity of the function relating pressure ratio to throttle position and compressor inlet temperature, an all-fluidic approach would be extremely complicated and not very accurate. It was therefore decided to generate the reference pressure-ratio signal by means of a three-dimensional cam.

The cam positions a static probe along the centre-line of a nozzle through which compressor air is being discharged to atmosphere. The static pressure at a point on the centre-line of a nozzle is a predictable fraction of the pressure at the inlet to the nozzle provided that shock waves are avoided. Thus a particular axial position set by the cam produces a pressure reference signal with which the compressor inlet pressure can be compared to produce, with proper

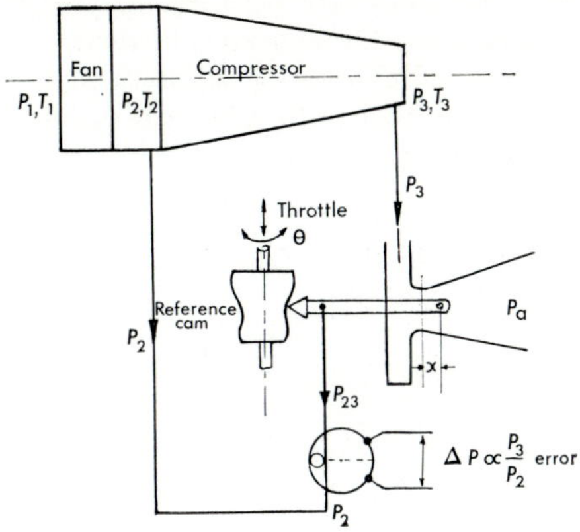

Figure 7.15 Pressure-ratio sensor.

scaling, a pressure-ratio error signal. The comparator is shown diagrammatically in fig. 7.15.

$$P_{23} = f_1\left(\frac{P_3}{P_a}, x\right)$$

and $\qquad x = f_2(\theta, T_2) \qquad \theta = \text{throttle position}$

Function f_2 is chosen so that—

$$\Delta P = P_{23} - P_2 = \left[\left(\frac{P_3}{P_2}\right)^* - \frac{P_3}{P_2}\right]$$

The asterisk denotes the reference value.

Pressures P_{23} and P_2 are compared in a proportional amplifier, the output (ΔP) from which is the pressure-ratio error signal. To complete the system two stages of proportional fluidic amplification are needed to operate the servo-valve as shown in fig. 7.16. Two capa-

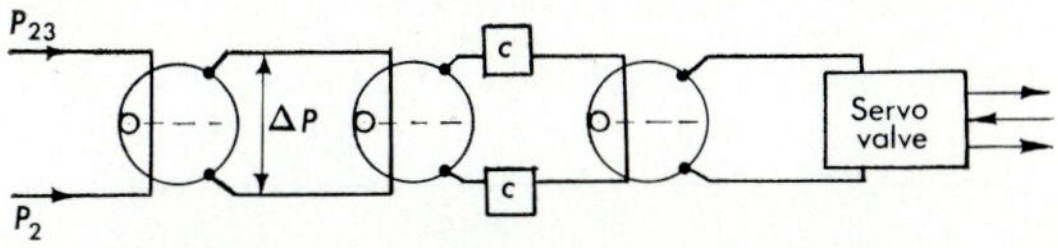

Figure 7.16 Pressure-ratio control circuit.

cities, C, are included in the amplifier to reduce the response from over 100 Hz to about 30 Hz. This is necessary to avoid instabilities by interaction with other control loops in the engine.

The now-common problems in fluidics of low signal-to-noise ratios and impedance matching were encountered during the development of the system. One of the most important was matching the probe to the amplifier, as a poor match gave rise to a noisy signal. This was overcome by back-pressuring the amplifier control ports to produce a slight back flow through the probe. This also had the advantage of preventing the ingestion of dirt.

In the final design the circuit had a supply which floated with the compressor discharge pressure to give operating characteristics unaffected by wide swings in the pressure being sensed.

7.3.2 *Speed loop*

The interesting feature in the mechanization of the speed loop is the use of fluidic carrier techniques in the comparator. Advantages

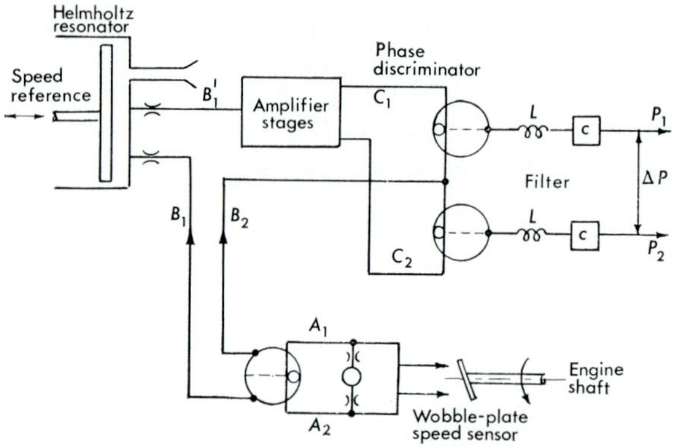

Figure 7.17 Speed-control circuit.

associated with a carrier system for this case include ease of generating a speed signal as a frequency, availability of an easily adjustable

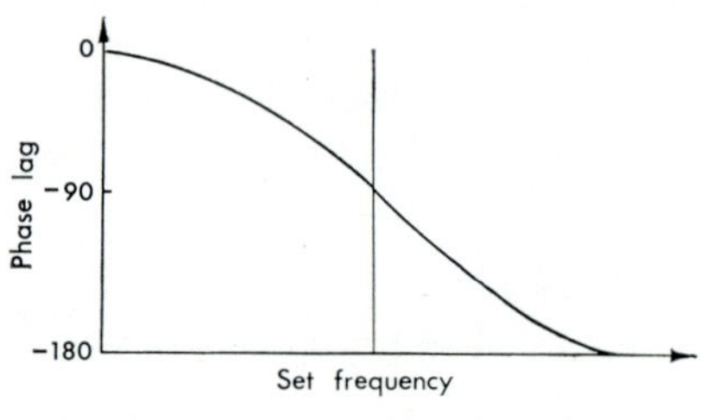

Figure 7.18 Resonator characteristic.

frequency reference, and tolerance of amplifier performance variations and minor leaks.

The circuit is shown in fig. 7.17. Engine speed is detected by a wobble-plate pick-off which emits a differential sinusoidal signal $(A_2 - A_1)$ of frequency proportional to engine speed. This signal is amplified to $(B_1 - B_2)$, and one half, B_2, is passed directly to the phase discriminator. The other half, B_1, 180° out of phase with B_2, is passed to the Helmholtz resonator. The resonator, together with some extra lag in the amplifier, has a phase characteristic of the form shown in fig. 7.18.

When the frequency of B_1 coincides with the natural frequency of the resonator a 90° phase-lag results. This is amplified, finally giving the push-pull signal $(C_1 - C_2)$ displaced 90° from signal B_2. The signals go to the phase discriminator, which consists of two fluidic

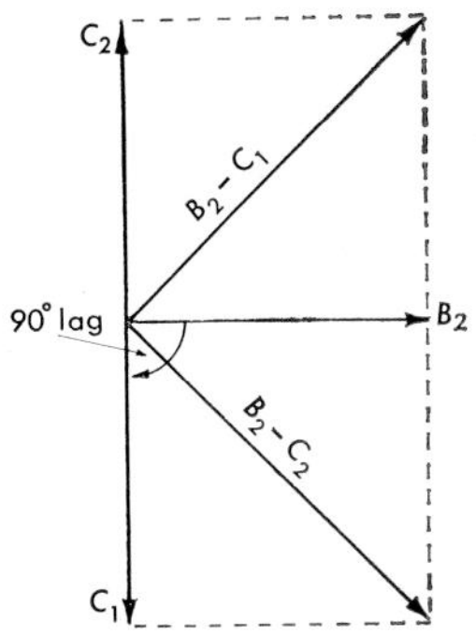

Figure 7.19 Phase diagram at correct speed.

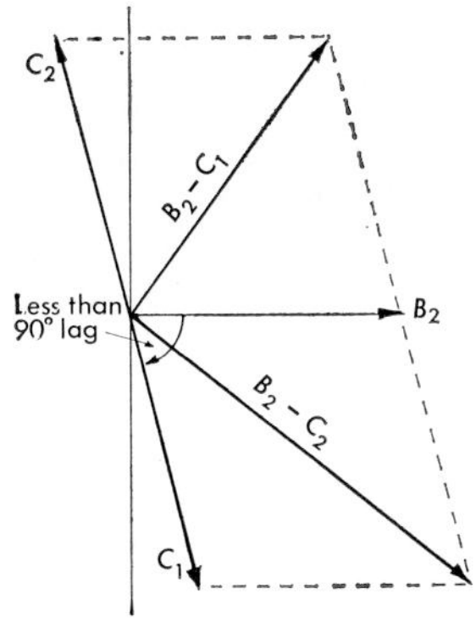

Figure 7.20 Phase diagram when speed is low.

rectifiers (proportional amplifiers with single central output). They are followed by filter circuits to extract the d.c. component of the rectified signal (the modulus). The phase relations of the signals are as shown in fig. 7.19.

The output speed-error signal $(\Delta P = P_1 - P_2)$ is the difference of the two moduli. If the speed is correct—

$$|B_2 - C_1| = |B_2 - C_2|$$

Hence $\Delta P = 0$ and there is no error signal. If the speed is below the reference frequency the phase lag decreases. Then the phase diagram will be as shown in fig. 7.20. Now—

$$|B_2 - C_2| > |B_2 - C_1|$$

so that $P_2 < P_1$ and a differential signal emerges. If the speed is above the reference the phase lag increases and an error signal ΔP of opposite sign $(P_2 > P_1)$ is given out.

Then differential pressure output operates a hydraulic servo-valve as described for the pressure-ratio loop.

7.3.3 *Temperature loop*

The temperature to be controlled is that of the turbine, control being effected by modulating fuel flow. The same carrier technique is used for the temperature comparator as for the speed loop. However, the method of producing the harmonic signal of frequency related to temperature is interesting and worthy of further notice.

The hot gas is sampled by means of a flow rake and mixed to give an even temperature. It is then passed through two fluidic oscillators whose frequencies are related to the square root of the absolute

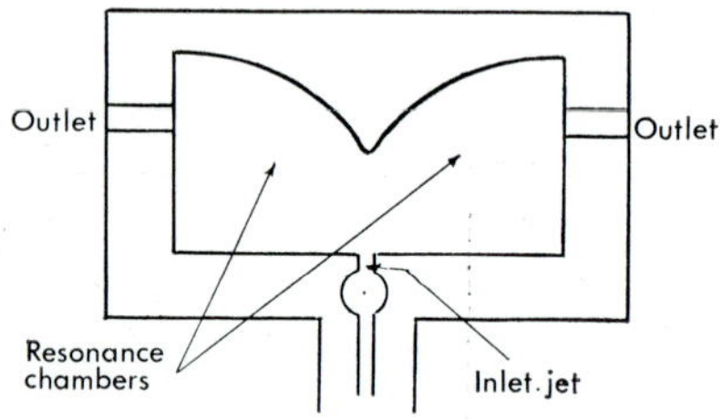

Figure 7.21 Temperature-sensitive oscillator.

temperature. The exact form of the oscillators used has not been disclosed, but it is probably similar to that shown in fig. 7.21. The inlet jet is directed against the sharp edge, which deflects it alternately into the resonance chambers. The frequency of oscillation, as stated

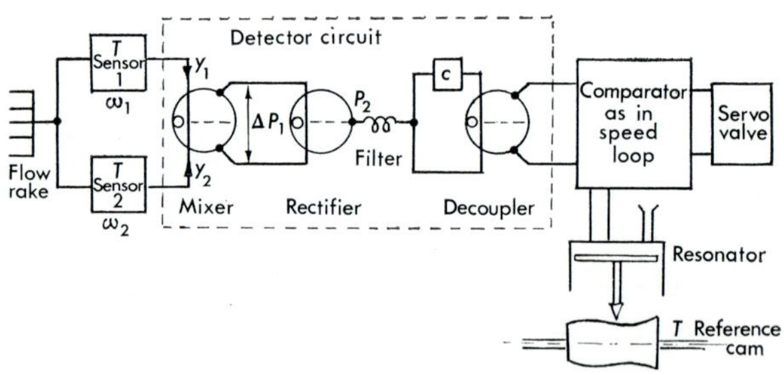

Figure 7.22 Temperature-control circuit.

above, is related to the absolute temperature. It is in the range 4 to 8 kHz. To reduce the frequency to a usable level, two slightly different frequencies from the two oscillators are mixed and subtracted in a beat-frequency detector circuit. The detector circuit consists of a proportional amplifier followed by a rectifier and a decoupler, as shown in fig. 7.22.

The signals from the sensors are fed into opposite control ports of the mixer/proportional amplifier. Assume signals of equal magnitude. Then—

$$y_1 = \sin \omega_1 t$$

$$y_2 = \sin \omega_2 t$$

The signal passed to the rectifier is—

$$\Delta P_1 = y_1 - y_2 = \sin \omega_1 t - \sin \omega_2 t$$

i.e.

$$\Delta P_1 = 2 \cos \tfrac{1}{2}(\omega_1 + \omega_2)t \sin \tfrac{1}{2}(\omega_1 - \omega_2)t$$

$$\tfrac{1}{2}(\omega_1 + \omega_2) = \text{carrier frequency} = \text{average frequency}$$

$$\tfrac{1}{2}(\omega_1 - \omega_2) = \text{beat frequency} = \tfrac{1}{2}\ \text{difference frequency}$$

The rectifier then takes the modulus (full-wave rectification)—

$$P_2 = 2\,|\sin \tfrac{1}{2}(\omega_1 - \omega_2)t \cos \tfrac{1}{2}(\omega_1 + \omega_2)t|$$

which gives a waveform of double the frequencies, that is $(\omega_1 - \omega_2)$ and $(\omega_1 + \omega_2)$, by inverting the negative half-cycles. The high-frequency signal is then filtered out, leaving the low-frequency signal. The frequency of the latter is related to the temperature. Finally the signal is reconstituted in differential form by the decoupler before being passed into the comparator circuit. The comparator uses a phase-sensitive detector circuit and puts out a 'd.c.' differential pressure, as already explained for the speed loop.

The signal (beat) frequency range is from 200 Hz at low temperatures to 500 Hz at 1100°C.

7.3.4 *Inlet guide valve control*

To optimize the economy and performance of a gas turbine while avoiding compressor surge it is required to proportion the angle of the compressor-inlet guide-vanes directly to $N/\sqrt{T}$ over a very small

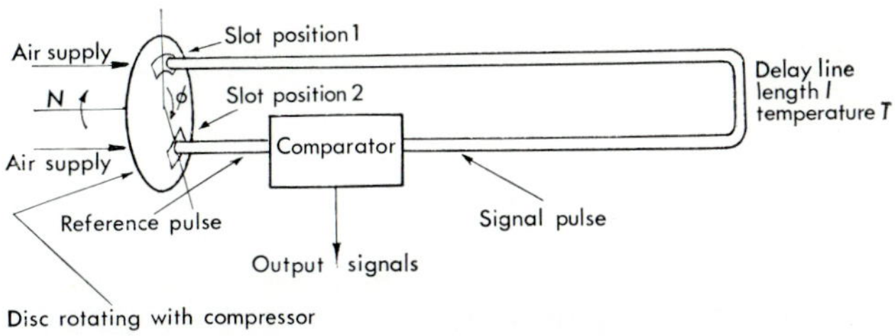

Figure 7.23 Basis of pulse-timing technique for $N/\sqrt{T}$.

range. Here N is the compressor speed and T is the absolute temperature at the inlet.

An interesting method for determining $N/\sqrt{T}$ has been developed by the Plessey Co. (*9*). It was specified at the design stage that the required accuracy of control was $\pm 2{\cdot}5\%$ of angle through only 10% of the range of $N/\sqrt{T}$. This required measurement of $N/\sqrt{T}$ to an accuracy of $\pm 0{\cdot}25\%$. Outside the specified range the vanes are held on limit stops. The required accuracy cannot be achieved with simple proportional fluidics so a pulse detection and timing technique is used.

The basis of the method is shown in fig. 7.23. The disc, geared to the compressor, carries a slot which sequentially uncovers two nozzles. When the slot is at position 1 a pulse enters the delay line and travels along it to the comparator in a time $t_1 \propto l/\sqrt{T}$ (since the velocity of the pulse is proportional to $\sqrt{T}$). When the disc has turned through an angle ϕ to position 2, a reference pulse is injected into the comparator with no delay.

Let t_2 be the time for the disc to turn through angle ϕ. Then $t_2 \propto \phi/N$. If the length of the delay line is such that both pulses arrive together then $t_2 = t_1$. That is—

$$\frac{\phi}{N} = \frac{l}{\sqrt{T}}$$

For fixed l—

$$\phi \propto \frac{N}{\sqrt{T}}$$

Thus a difference between the times taken by the two pulses to reach the comparator can be used as a correction signal to vary the guide-vane angles and angle ϕ to suit the existing value of $N/\sqrt{T}$.

An improvement is shown in fig. 7.24. The delay line is flushed

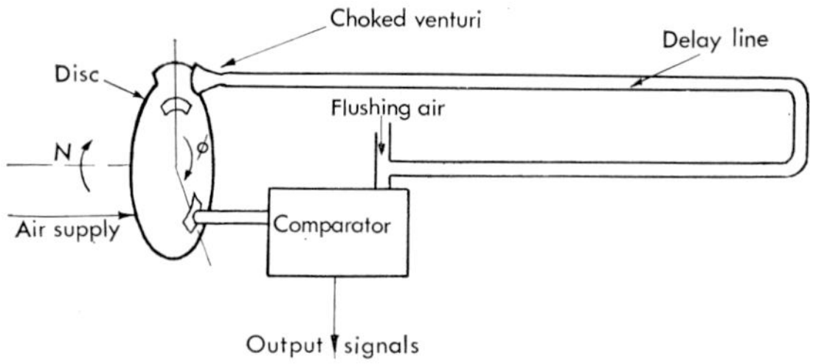

Figure 7.24 Improved pulse times.

with air of the correct temperature T and the signal pulse is produced by momentarily blocking the outlet of a choked venturi (choked to keep the Mach number in line constant)* at the end of the delay line. This method has the advantages of ensuring that the temperature in the delay line is correct and of preventing the ingestion of dirt.

The comparator provides a three-state output to drive the servo-system which controls the guide-vane angle θ and the angle ϕ between pick-up points. If the two pulses arrive simultaneously the comparator gives zero output and the servo-mechanism does not move.

If the delay line pulse arrives first an output is generated. This causes the servo-mechanism to reduce ϕ and vary θ accordingly until

* A *venturi* is a nozzle which is convergent at entry and divergent at exit. It is said to be *choked* when the fluid velocity at the throat reaches the local speed of sound and the downstream pressure no longer affects conditions upstream. (Cf. Section 5.3.)

two further pulses arrive simultaneously. Then the servo-system stops at the new position in which the value of θ corresponds to the existing value of $N/\sqrt{T}$.

If the reference pulse arrives first the opposite output is generated

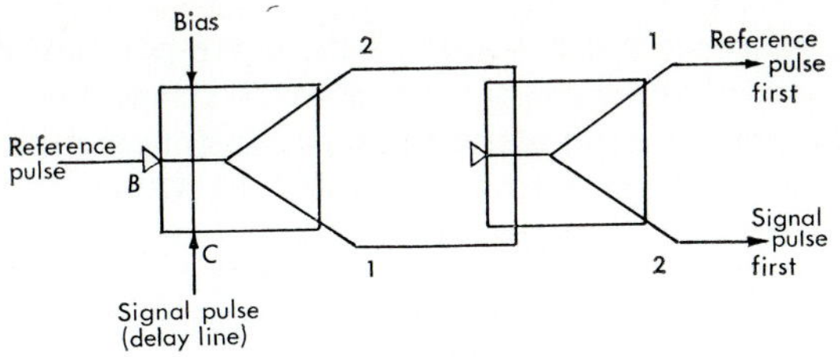

Figure 7.25 Comparator unit.

by the comparator and the servo-mechanism increases ϕ and varies θ until once again two pulses arrive simultaneously.

The basic unit of the comparator is a modified form of an ultra-sensitive amplifier (*10*) and is shown in fig. 7.25. If a reference pulse arrives first the bias directs it out of output 1. If the signal pulse

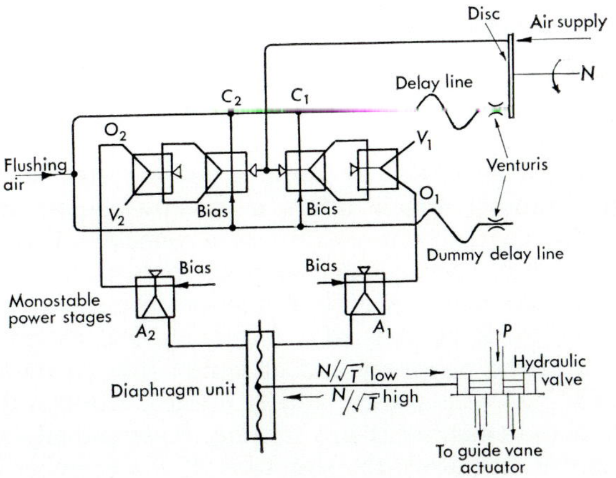

Figure 7.26 Simple complete system.

arrives first it backs off the bias and the reference pulse is directed out of output 2. Pressure levels are arranged so that arrival of the signal pulse after the reference pulse will not alter the output.

To get the required behaviour from the system two comparator units are used as shown in fig. 7.26. In this case the bias is provided by a dummy line flushed by the same air supply as is the delay line. The dummy delay line cancels out noise and supply pressure variation. The connexions to the delay line are a small distance apart but the reference pulse connexion is common so that both units receive reference pulses at the same time. The comparators each have one output vented and one driving a monostable power amplifier. The power amplifiers drive a diaphragm unit connected to a servo-valve. The valve controls the guide-vane actuator, which also varies the angle ϕ.

When a reference pulse is applied the corresponding signal pulse may be in one of three positions relative to connexions C_1 and C_2. If the signal pulse has not arrived at C_1 or C_2 (in other words $N/\sqrt{T}$ is high) the bias causes outputs at O_1 and V_2 so that amplifier A_1 is switched on to drive the diaphragm and servo-valve from right to left (see fig. 7.26). If the signal pulse has arrived at C_1 but not C_2 the outputs will be at V_1 and V_2 so that the diaphragm unit is depressurized and the servo-valve is closed. This is the case $t_1 = t_2$. If the signal pulse has arrived at both C_1 and C_2 (in other words $N/\sqrt{T}$ is low) when the reference pulse arrives, the outputs will be at V_1 and O_2 so that the servo-valve is driven from left to right.

In a fully operational system various interlock and override elements are needed. They are omitted here for clarity.

7.4 The vortex angular-rate sensor

An important component which has been mentioned in Section 1.8, and several times in this chapter in connexion with flight systems, is the rate sensor.

The normal form of this device in a fluidic system is the vortex angular-rate sensor, which offers a no-moving-part means for measuring the angular rate of turn of a vehicle and replaces the highly refined, but expensive, gyroscopes at present universally used.

The vortex rate-sensor is a close relation of the vortex amplifier (see Section 1.8). It consists of a flat cylindrical chamber with a porous wall at the periphery and an outlet tube at the centre, as shown in fig. 7.27. Gas passes radially inwards through the porous wall, and, if the chamber is not turning, flows radially across the vortex chamber and down the sink tube. If the chamber is rotated with angular velocity ω, a tangential component of velocity is imparted to the fluid as it passes through the porous wall, making it spiral across the vortex chamber. The angular momentum of the flow is conserved across the chamber so that the tangential velocity component increases in inverse proportion to the radius. The flow finally passes down the sink tube where the helix angle, ϕ, of flow is proportional (within limits) to the rate of rotation of the chamber [see fig. 7.28(a)].

The flow angle may be measured with different forms of pick-off, one simple design for which is shown in fig. 7.28. This pick-off consists of two small tubes, each with a blanked end and a small sensing hole. The tubes are positioned across the sink tube so that one sensing hole lies on either side of and close to the centre-line of the sink tube and at an angle θ as shown in fig. 7.28(b). In this position the rotating flow, which has very high tangential velocity near the centre, is incident on one sensing hole and away from the other. The result

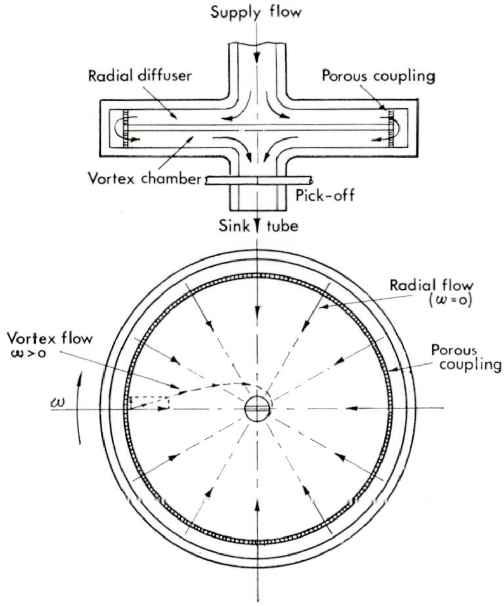

Figure 7.27 Vortex angular rate sensor.

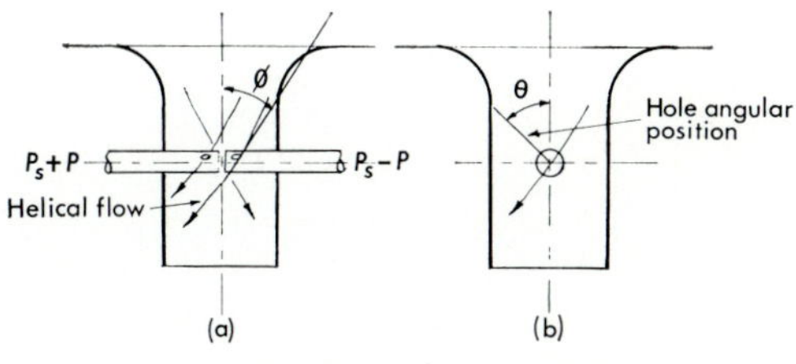

Figure 7.28 Sink tube and pick-off.

is that the 'incident' sensing hole measures the static pressure plus a fraction of the dynamic pressure $(P_s + P)$ and the other hole measures the static pressure minus a fraction of the dynamic pressure $(P_s - P)$. The pressure difference $(P_s + P) - (P_s - P) = 2P$ is then proportional to the rate of rotation. Very little flow can be drawn from the pick-offs without reducing the gain [pressure/(rad/s)], so the rate sensor is usually followed by a high-input-impedance, fluidic, proportional amplifier to buffer it as in the system of fig. 7.3.

Typical performance figures for a 150 mm diameter chamber are:

Gain 40 $(N/m^2)/(^\circ/s)$ and a transportation lag of 10 ms
Output threshold 0·5 $^\circ/s$

A problem with the design is that sensitivity falls with diameter, so installation, in very small missiles for instance, becomes a problem too. However, many vortex rate-sensors have been made, tested and flown successfully and the device seems to have a significant contribution to make to aerospace development.

Acknowledgement

I acknowledge with thanks permission by Reuben H. Donnelley Corporation to reproduce text and diagrams from their journal *Control Engineering*, the copyright in which is their property.

8

TRENDS AND PROSPECTS

B. J. Cooper and D. L. Hughes
English Electric Co. Ltd

In this final chapter we shall cover the trends and prospects for fluidics in terms of the elements themselves, their applications, and the markets for them. The chapter is subdivided into the following main sections: 1. Development of elements, 2. Manufacturing techniques—system developments, 3. Applications, and 4. Future markets.

8.1 Development of elements

8.1.1 *Logic elements*

Though appearing in many guises, most of the pure fluidic elements in current use are essentially either turbulence amplifiers or wall-attachment devices. Each type has its own advantages and disadvantages. Turbulence amplifiers offer good isolation of inputs and moderate power consumption, but they are inherently limited to *nor* operation, have a rather low switching speed, and are prone to acoustic disturbance. Wall-attachment devices are faster, have higher logic power, and are insensitive to noise and vibration, but they tend to consume more power and may suffer from signal interactions.

The trend in developing these devices has lain in reducing the disadvantages by attention to the detail design of existing elements. This has been followed by more radical changes, sometimes involving new operating principles, in a search for a second generation of devices.

Progress with turbulence amplifiers has followed this pattern closely. The original, rather large, tubular devices were superseded by small, planar elements. Higher packaging densities were then obtained, with rather better static performance and higher switching speed. This in turn was followed by an increase in the number of input ports to improve utility. At about the same time Rolland (*1*) reported a modification of the basic element—incorporation of a wall-attachment phase to improve performance and increase logic power. The future trend is likely to be continued reduction in size leading to a miniature multi-element block with built-in acoustic shielding.

Wall-attachment elements have undergone less radical changes and most of the improvements have been obtained by detailed design of the geometry, particularly the vents, to isolate inputs and reduce load sensitivity. Early predictions (*2*, fig. 8.1) that these elements would have a well defined lower Reynolds number, leading to increased size at lower operating pressure in low-power-consumption

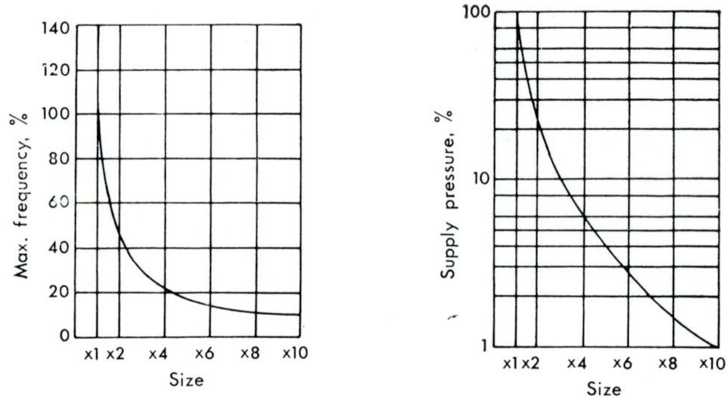

Figure 8.1 Effects of element size on maximum operating frequency and supply pressure when Reynolds Number constant and minimal (*2*).

elements, have not been borne out entirely. Highly miniaturized elements, in some cases using trigger points to ensure turbulence in an essentially laminar flow, have been operated at gauge pressures well below the 0·2 lbf/in^2 (1·4 kN/m^2) normally used in turbulence amplifiers. These small elements have the advantage of a higher operating speed—it can be as great as 1·5 kHz, compared with 200 to 30 Hz for normal turbulence amplifiers. They generally have lower recovery pressure and a smaller fan-out than turbulence amplifiers, but Sarpkaya and Kirshner (for example) have reported details of a large-model curved-wall attachment amplifier which has outstanding performance characteristics in all these respects (*3*). This particular design is difficult to miniaturize, but progress is likely with future devices, limited only by increasing viscous losses as size is decreased.

Very small elements tend to become more sensitive to accumulated deposits in the working region and are prone to blockage by stray fragments in the circuits. One approach to this problem is to package the elements with vent-protection filters, and in some cases individual coarse screens in all the signal and power ports. This is satisfactory in more sophisticated applications, in which some additional cost can be tolerated, and where minimized power consumption with high speed of response are the major considerations.

However, many applications can be envisaged, particularly in the technologically less advanced industries, in which the main requirements are tolerance of poor air supplies and lack of cleanliness during assembly. Here it seems probable that elements considerably larger than the early ones will be used. They will have relatively slow response unless supplied at high pressure, but they should be capable of swallowing without ill effects the bits and pieces which can be found in badly made pneumatic circuits.

The probable short-term outcome will therefore be two distinct ranges of logic element, one highly miniaturized, with sophisticated packaging for optimum performance at moderate cost, the other much larger, possibly with lower performance but with great tolerance of misuse. To be successful both types will have to be insensitive to driving and loading conditions. Beyond this there will be a continued effort to increase the maximum switching frequency to 5 or 10 kHz by using newer designs such as the edge-tone amplifier (*4*), and to reduce power consumption even further. It is important to note that, while both these objectives can be achieved within a logic circuit by miniaturization, this can lead to a reduction in overall operating speed in some instances, if interface devices are not also developed to accept the low-flow–low-pressure signals without loss of response speed.

The main moving-part elements which are normally considered in connexion with fluidic elements are miniature diaphragm relays, foil elements and biased-spring devices. Diaphragm relays have the advantage of working at pressure levels up to 20 lbf/in^2 (138 kN/m^2) with sufficient flow capacity to operate small output devices directly. They also require flow only transiently when operative, and in circuits with low switching activity this can lead to substantial savings in power. They are rather slower than pure fluidic devices and are not likely to be packable at very high densities. Foil elements, on the other hand, normally work at low pressures and can be densely packaged. Captive foils are probably subject to fatigue failure while free foils seem prone to sticking if contaminated air is used. Development in these cases has been towards size reduction and simplification to overcome the disadvantages and obtain the highest possible packing density (*5*). Biased-spring devices offer many advantages in ease of connexion, higher output pressures, etc., but are also likely to have a life limited by fatigue and contamination of the spring.

In general, moving-part logic devices have advantages which make them attractive for many current applications but seem to be relatively limited in development potential. It is therefore to be expected that they will be overtaken by improved pure fluid devices, with superior life and operating speed, in all but the specialized applications where their flow shut-off capabilities are essential.

One inevitable trend is towards the use of planar foil diodes to improve the fan-in of wall-attachment elements (*6*). This often leads to considerable overall simplification of the circuits, which can to some extent counterbalance the slight penalty in speed of response. Passive pure fluid diodes are unlikely to achieve comparable characteristics, and this type of hybrid device could become more widely used if it proved to be free from vibration and contamination troubles.

8.1.2 *Analogue elements*

Three major types of analogue element are in use. They are the beam-deflexion amplifier, the impact modulator and the vortex amplifier.

In the first type a jet is deflected between two receivers by the momentum or static pressure of one or more cross jets. This type can achieve a static pressure gain of about ten in a single stage, with quite good signal-to-noise ratios. Development in this case has proceeded mainly by detail modifications to improve these characteristics, though some attempts have been made to introduce major features such as positive feedback in the vents to improve gain. Kelly and Shinn have also reported stratification of the element into a number of parallel paths to reduce noise generation (*7*).

In the impact modulator the radial flow pattern produced by the collision of two opposed jets is monitored by a receiver. The pattern is moved relative to the receiver either by varying the strength of the jets or by deflecting one of them sideways by means of a third jet. This latter type can give pressure gains of 200 and is therefore usable as a crude operational amplifier.

Trends in both cases are likely to lead to improvements in gain, stability and signal-to-noise ratio as a result of detail changes rather than radical modifications. The tendency here will be towards the development of packaged operational amplifiers of sufficiently high standard to be used as analogue computing elements in pneumatic control systems.

One approach to the problem of d.c. amplification at low power levels has been to employ a.c. carrier systems using either amplitude- or frequency-modulation techniques. Much has been published over the past few years on a.c. fluidic systems (*8–13*). Broadly speaking the advantages of a.c. over d.c. in analogue systems are similar to those with electronics, i.e. improved signal-to-noise ratio, particularly in low-level amplifiers, elimination of drift effects due to power supply changes, temperature etc., with a consequent increase in accuracy.

Up to the time of writing the various systems described have been experimental. In general, analogue fluidics is less advanced than digital. Several of the leading companies in the U.S.A. are reported to be active in this field. For example, General Electric has demonstrated a steam-turbine-governing system, and is also using the technique for temperature control. Honeywell is using similar systems for aircraft engine control.

The rate of application of a.c. systems in the future will depend upon the development of the supporting technology. The analysis and design of the systems require a composite technology incorporating acoustics as well as heat and mass transfer, and this is not at present sufficiently well developed.

Other factors which may significantly affect the future for a.c. systems are further developments in analogue circuitry. The analogy between fluidics and electronics is not a very exact one. For example, there are no inherent temperature effects in fluidic amplifiers that correspond with those in transistor amplifiers. It is not necessarily true that solutions developed for electronic circuitry will automatically be the optimum solution for fluidic circuitry. A.c. fluidics may well fall into this category when the problems in analogue circuits are more clearly defined and understood.

Another analogue element in general use is the vortex amplifier. This is essentially a flow-control valve in which a high-pressure signal can control the flow out of a power port by swirling the flow in a chamber. The future development of this type of element is likely to be concentrated on the improvement of turn-down ratio with lower signal levels, and to permit use of the amplifier as a load-controlling valve which is free from hysteresis and which can handle heavily contaminated fluids if necessary.

8.1.3 *Element standardization*

From the above discussion of trends it will be obvious that, up to the time of writing, there has been little opportunity for standardization of element functions or dimensions.

Attempts have been made to introduce standardized 'functional' and 'operating principle' symbols (*14*, *15*), and to a lesser extent suggestions for common nomenclature and data presentation have been made. Many manufacturers' data sheets, however, give insufficient information to permit any form of circuit design other than trial and error. This is a situation which will have to be remedied as one of the suggested standards comes into general use and manufacturers extend their ranges of components.

It is unlikely that any particular element function will predominate to the exclusion of all others, though there are some arguments in favour of using solely *nor* logic when ease of manifolding in standardized mounting systems is considered.

One trend probably will be toward the gradual disappearance of the present multiplicity of connectors in favour of a few ranges of standardized port grids to which discrete elements from all manufacturers may be mated. Progress in this direction would be assisted by the appearance of preferred working-pressure ranges and nozzle sizes to ensure some degree of matching between elements from difference sources.

8.1.4 *New operating principles*

An examination of many of the logic devices claimed to work on new principles shows that they are actually variations on the basic wall-attachment or laminar-turbulent transition devices. There are some exceptions, however, such as the device in which positive-feedback momentum interaction determines the output state. Another typical example being investigated for high-speed applications is the edge-tone amplifier, in which vortices are shed from an edge to cause a device to oscillate continuously between the outputs until a signal is applied causing it to stay hard over on one side. It is claimed that, if the basic edge-tone frequency is made as high as possible, the time required to deflect the output fully is greatly reduced in comparison with that for normal wall-attachment devices. A further possible switch is the double-throat supersonic nozzle, in which the shock wave may be moved between two stable positions by a signal. The output in this case is the pressure rise across the shock wave.

Most of the reported work on analogue elements deals with attempts to alter the characteristics of conventional amplifiers by internal feedback. In some cases this has amounted to use of limited wall attachment to augment the signal. In others part of the main jet has been bled off and fed back to obtain the same effect.

Many fluid flow phenomena show stable and unstable states and are potentially usable in fluidic devices. It is probable that most of them have been investigated already, and it is difficult to predict any major change in operating principle for second-generation devices.

One probable future development is extension of the use of liquids rather than gases. Taft (*16*) has shown this to be feasible with conventional elements operating at appropriate Reynolds numbers but has pointed out problems of high power consumption, low operating speed and cavitation due to liberation of dissolved air. It is likely that special elements will be required for liquid operation. Their design is being actively pursued.

A further possible development is the exploitation of the Winslow electro-viscosity effect (*17*) for control of fluid flow. The fluid contains fine solid particles in suspension and control is exercised by means of a static electric field. The particles may cause problems with some of the auxiliary equipment, such as pumps, but at least in concept this is probably one of the most useful new analogue control methods currently under investigation.

8.2 Manufacturing techniques—system developments

As the various manufacturing processes for fluidic elements and systems develop, it seems certain that plastics injection moulding will play a major part in all applications for normal environmental conditions. This method gives the lowest cost in large-quantity production and will obviously result in significant price reductions when production quantities rise as the market expands.

The movement in electronics over the past fifteen years, from wired discrete components through printed-circuit mounted components to integrated circuits, is already discernible in fluidics. The initial stage of piped-up individual components—the 'spaghetti era'—is already passing, and we are moving towards the use of individual components manifold-mounted on some form of circuit plate.

The next stage is likely to be the one-piece multi-element circuit. This also will probably follow the electronic trend, and, except for certain widely used circuits such as counters, shift registers etc., circuit plates will consist of collections of separate elements—logic gates, flip-flops etc.—with all their connexions brought out, and the network formed externally by a circuit board. Such multi-element circuit plates are already being produced in photo-sensitive glass-ceramic and epoxy resin. Precision tools are already being developed to produce plates in large quantity by injection moulding in plastics.

At the time of writing fluidic modules are interconnected by piping, but the trend is again following electronic precedent in that 'plug-in' modules are beginning to appear. The duty of fluidic plugs and sockets is every bit as onerous as that of their electronic counterparts, and development effort will be necessary to achieve the required life and reliability.

For applications in extreme environments the most likely methods of manufacture are either (i) photo-etching or moulding in glass, or (ii) etching, chemical machining or spark machining in metal.

One feature of fluidic elements which has no counterpart in electronics is the need for vent protection. When the early fluidic circuits were used in industrial atmospheres it rapidly became apparent that it was not sufficient to filter the air supply. Some filtration was also necessary on the vents of the element, which under certain operating conditions entrain air from the atmosphere and hence draw in dirt. The immediate solution in many cases has been to enclose the system in a sealed box slightly above atmospheric pressure and connected to the outside atmosphere *via* some form of filter. With separate modules this solution is not ideal as in many cases the modules are required to work without any case or enclosure. Modules must have built-in protection. This is provided either by individual filters on each element vent or by vents discharging into an enclosed space. Some form of the latter system is likely to become standard in the future, permitting as it does the vent flows to be collected and piped if required. This is normally essential when using working fluids such as natural gas and similar explosive or noxious gases.

8.3 Applications

In order to look into the future it is worth while summarizing the main advantages and disadvantages which fluidics are likely to show when compared with other techniques.

8.3.1 *Advantages*

There are four main areas of advantage. They have to do with questions of environment, the compatibility of fluidic devices with existing working fluids, the availability of simple, reliable sensors for many applications, and (in certain circumstances) cost.

The environmental factors include both the difficult environments —high temperature, radiation etc.—and the hazardous atmospheres in which fluidic circuits are inherently safe. The ability conferred by fluidics to cope with extreme environmental conditions has opened up many new areas to control for which previously there were no suitable components. Typical of these are nuclear-reactor controls (see Chapter 5), which may now be housed in the high-temperature, high-radiation areas, in many cases within the reactor pressure vessel, and aircraft engine controls (see Chapter 7).

The compatibility factors include both systems in which conventional working fluids are already available, e.g. compressed air or oil under pressure, and also systems in which there are other pressurized fluids which may be used to power a fluidic system. These include steam, natural gas, boiler feedwater, and a range of gases and liquids in chemical processes.

Many of the input sensors required by fluidic circuits are extremely simple and robust. Typically, movement and proximity may be sensed by simple, interruptible jets or back-pressure sensors. Many variables such as fluid flow and pressure lend themselves to fluidic circuitry, whereas electronic systems require complex and expensive transducers.

When considering cost it is very difficult to make any definite predictions owing to the rate at which the prices of alternative systems (electronic ones) are currently falling, and also because the present prices of fluidic components are highly artificial. In general, however, it may be assumed that fluidic systems are unlikely to have any significant cost advantage over alternatives unless (a) there is an environmental condition that requires expensive packaging and protection for the alternative system, or (b) the system interfaces or sensors and output transducers favour fluidics.

8.3.2 *Disadvantages*

The major disadvantages of fluidics are the high power consumption and the slowness of response.

Power consumption is likely to be reduced by development of elements (see Section 8.1.1), but it is unlikely that fluidics will ever be comparable with electronics on this score. There is also inherent conflict between low power consumption and high reliability in fluidic elements owing to the reduction in passage sizes which reduced power consumption entails.

It is of interest in this respect to note that low power consumption is emphasized in fluidics for quite different reasons than in electronics. In electronics a large proportion of all the power consumed appears as heat, and the main objective in reducing power consumption is the reduction of this heat to permit higher packing densities and greater reliability. Provision of the necessary electrical power is, in general, no problem in mains-powered equipment. In fluidics the major problem is the supply of the necessary working fluid. Even when a 'mains' compressed-air supply is available, it is often of limited capacity and a large fluidic system can easily overload it.

Fluidic and electronic manifestations of portable and self-contained systems are more similar, though at the time of writing fluidic circuitry and portable power supplies compare very unfavourably with very low-power-consumption modern electronic circuitry and high-efficiency batteries. The problems of power consumption, even with the miniature elements now being predicted, have already led I.B.M. to study the foil element (*5*). Although not strictly fluidic, since it contains a moving foil, this element appears very attractive for systems like computers and data processors where very many elements are required. In such applications foil elements have the advantages of complete shut-off—i.e. zero air consumption in the *off* condition—and considerably smaller size than conventional fluidic elements, permitting greater packing densities.

The speed of response of fluidic circuits is set by two factors. One is the operating time of the individual elements. A typical switching time is 1 ms, and future miniaturized elements are likely to be faster (see Section 8.1.1). The second factor affecting response speed is transit time. The minimum transit time of a pressure signal is fixed by the velocity of sound in the operating fluid. In air at normal pressures and temperatures it is typically about 1 ms/ft (3 ms/m).

This is obviously a fundamental limitation. It means that, however much the speed of response of individual elements may be raised, the overall system response will be largely determined by the physical size of the system. So fluidic systems have a very great disadvantage where signal transmission over great distances is required, and it is unlikely that fluidics will find many such applications. Although speed of operation may be increased in small, compact systems, there will always be a severe restriction on the use of fluidics where fast response is required.

8.3.3 *Particular cases point the way*

Having briefly considered the advantages and disadvantages of fluidics, let us consider likely future trends in certain selected areas of application.

Fluidics is just beginning to be taken seriously. It is no longer regarded as a 'gimmick' but is being considered in earnest by designers as a possible control tool. In the present state of the fluidics industry many applications fall down because of the lack of a comprehensive range of back-up hardware, but there is every sign that these gaps are being filled, and new components are appearing at an increasing rate. Advantage is being taken of electronics experience. Fluidic components are being designed to mount in electronic racking, and components such as switches, indicators etc. closely follow the appearance of their electronic equivalents.

As fluidics markets grow, applications will obviously be found right across the industrial front, so it is difficult to highlight any particular area. The areas covered below are obviously not exclusive, nor do they necessarily represent the major markets. They have been chosen to cover diverse applications of technical interest.

8.3.4 *Gas-turbine control*

The first work in this field was carried out in the U.S.A. by Honeywell and General Electric and was concerned with the control of aircraft gas-turbines. In 1960 the advent of high-speed aircraft, with the resulting high operating temperatures for engine control components, was already causing great concern. Neither electronic nor hydromechanical controls could operate at the required temperatures and the early developments in fluidics occurred at the ideal time. In 1963 Honeywell and General Electric were awarded development contracts for fluidic controls, and the work carried out on these contracts has been widely reported. Systems have operated up to temperatures of 1200°F (650°C). Experience with these controls was so successful that they are now being applied to industrial gas-turbines, and it is claimed that the accuracy of the temperature control is such that the turbines may be safely up-rated.

It is of interest that this initial application utilizes most of the advantages of fluidics listed above, i.e. insensitivity to severe environments and compatibility with existing working fluids—the controls are powered by air from the engine. Also, rugged and fast-response sensors are used to measure engine variables.

For the future it is claimed that, in the gas-turbine control field, fluidics will overcome the problems of high cost which obstruct expansion of the gas-turbine market at the time of writing. Fluidic controls are claimed to be competitive with conventional controls on industrial gas-turbines, and show additional advantages in their long life, insensitivity to fuel contamination and their ability to give close temperature control during start-up.

One problem which has not yet been solved is that of sensing engine surge. Present controls compute this as a function of turbine inlet temperature and speed (*18*).

8.3.5 *Automotive engineering*

A spate of patent applications is appearing in connexion with the use of fluidics in carburation. This is a natural area for fluidics, involving as it does the control of fluid flow and mixing. Some early experiments have already demonstrated feasibility (*19*), but they have been very basic and have retained much of the conventional carburettor.

The present trend is towards 'computerizing' carburation, making the mixture strength respond to many additional variables such as engine speed, inlet manifold vacuum, torque and temperature. At the time of writing this is done electronically, but as fluidic technique develops it could show many advantages here, including greater reliability and lower cost.

Fluidic controls have also been suggested for windscreen wipers and power steering (*20*).

8.3.6 *Air-conditioning systems*

'Power' fluidic elements as diverting and mixing valves for air-conditioning systems offer many advantages. The absence of moving parts gives simpler and more reliable control. Few developments in this field have been reported, though some patents have appeared both in the U.K. and U.S.A. and Bjornsen (*21*) has described a pneumatic controller-receiver for environmental control installations based on impact modulator amplifiers with a moving-part output amplifier.

8.3.7 *Steel industry*

Although the present trend in this industry is towards centralized computer control, such control requires local monitoring and actuation loops, often operating in severe environmental conditions.

Typical are detection systems for hot steel strip and rod entering rolling mills, edge-sensing and strip-positioning controls. Final actuation in such systems is normally pneumatic. This, combined with the severe environment, high temperature and dirty atmosphere, together with the low cost and robustness of air-jet sensing systems, makes fluidic control a natural choice. Fluidics are already beginning to find applications in these areas, and the trend is certain to continue.

8.3.8 *Automatic process control*

The situation here is very similar to that in the steel industry. The trend is towards central computer control with local controllers. The use of digital fluidic controllers in conjunction with a central computer has been proposed. In this system the fluidic controllers handle much of the lower-speed logic so that a simpler, cheaper, central computer can be used (*22*).

8.3.9 *Textile industry*

The textile industry, in general, is one of the traditional users of pneumatics. Electronic systems have barely penetrated the industry yet, and there abounds a somewhat illogical but nevertheless very real preference for pneumatics and mechanics to electronics. In such an environment one would expect fluidics to gain rapid acceptance. All the signs are that this is happening, and that the textile industry will become one of the major users of fluidics. Air supplies are already available on much textile machinery, and air is used for the transport of yarns and for clearing broken yarn-ends etc. Fluidic systems for tension control in textile yarns have already been reported (*23*), and systems for broken-end detection are under development.

More sophisticated control, for example control of pattern in knitting machines, is still largely exercised by mechanical systems in conjunction with mechanically read punched-card or Jacquard systems, and is likely to offer some applications for fluidics on slow machines. The speeds of response required to handle the input data on fast machines are at the limit of present fluidic technology.

8.3.9 *Nuclear reactor control*

With the advent of fluidics it is possible to exert control in many areas of nuclear reactors where control was not previously possible owing to the high levels of radiation and temperature. Because such areas are inaccessible once the reactor is commissioned, only the very simplest and most reliable moving-part systems could be used within the reactor pressure vessel. Fluidics now offers the possibility of incorporating non-moving-part controls, actuated externally by pressure or flow signals, to control directly such variables as fuel-channel flow-rates and temperatures. Such controls offer significant improvements in reactor output as all fuel channels may be operated under their optimum conditions.

8.3.10 *Domestic applications*

Many domestic applications for fluidics have been proposed but the only ones we know to be in production are a fluidic lawn sprinkler and a variable shower spray. As water under pressure is readily available domestically it is reasonable to suppose that water-powered fluidics will appear in the future in a range of domestic appliances. The obvious applications are controls for dish-washers and automatic washing machines, and patent applications covering such systems have already been published. Other possible domestic applications include mixing controls for hot and cold water, water-powered toothbrushes (*20*), ornamental fountain controls and vacuum cleaners.

8.4 Future market

There have been many estimates made of the future market for fluidics. They range from £5m to £300m p.a. in 1975. Looking back over past market predictions for fluidics we see one thing clearly. The time scales in all cases have been extremely optimistic. Although there is no doubt that fluidics is here to stay and is growing, the rates of growth are much less than earlier predictions would suggest.

There are many reasons for this slow acceptance, and the major one is the reluctance of industry to use any new, unproven product. During the past two years many equipment manufacturers have been evaluating fluidics over a range of products. Large orders will not be placed, however, until these evaluation trials are successfully concluded, and with a completely new technique such as fluidics the trials are necessarily long-term.

The penetration of fluidics has also been restricted by the lack of much of the associated peripheral hardware—fittings, indicators, power supplies, etc. Although usable fluidic elements have been on the market for several years, it has been only recently that much of the associated hardware has become available. In the early days many standard pneumatic components and fittings were used, but now a range of hardware designed specifically for fluidic applications is appearing. The items are more compact, and owe more to electronics than pneumatics in many aspects such as packaging, appearance, etc.

The fluidics market to date has largely been in components sold on a 'do it yourself' basis for one-off special systems. The main markets, in systems produced on a quantity basis, will only appear as manufacturers begin to fit fluidic controls etc. as original equipment. This stage is only just being reached.

Probably the most reliable assessment of the future market potential is given in the recent survey commissioned by the Ministry of Technology (*24*). This assesses the U.K. market for fluidics as being between £5m and £15m in 1975, with a European market about three times this, and forecasts that the major growth areas will be in the material handling and process industries. The table opposite is taken from this report, and shows the predicted composition of future markets.

		1968–1970	1970–1972	1972–1975	1975–1980	GROUP
		SHORT-TERM		LONG-TERM		
Aerospace	E M	Large Very small	Large Small	Large Medium	Large Medium	b
Road and rail	E M	Very small Zero	Small Very small	Small Very small	Medium Small	b
Marine	E M	Medium Zero	Medium Very small	Medium Very small	Medium–large Medium	b, a
Machine tool	E M	Large Small	Medium Small–medium	Medium Medium	Medium Medium–large	c, b
Material handling etc.	E M	Medium Small	Medium Medium	Small–medium Large	Small–medium Large	c
Domestic	E M	Very small Zero	Small Very small	Small Small	Small–medium Medium	a
Nuclear	E M	Small Very small	Small–medium Very small	Small–medium Small	Medium Small	b
Medical	E M	Very small Zero	Very small Zero	Very small Zero	Small Small	a, b
Agricultural	E M	Very small Zero	Very small Zero	Small Very small	Small Small	c
Office equipment	E M	Very small Zero	Very small Zero	Very small Zero	Very small Very small	a
Miscellaneous	E M	Medium Small	Medium Medium	Small–medium Large	Small–medium Large	c
Chemical	E M	Very small Very small	Small Small	Medium Medium	Medium Medium–large	a, b, c
Gas	E M	Very small Zero	Small Very small	Small–medium Small	Medium Medium	a, b, c
Petrol	E M	Very small Very small	Small Very small	Medium Small–medium	Medium Medium	a, b, c
Textile	E M	Small–medium Small	Small–medium Small–medium	Small–medium Medium	Small Medium–large	a, b, c
Iron and steel	E M	Very small Very small	Small Very small	Medium Small	Medium Medium	c
Food and drink	E M	Very small Zero	Very small Very small	Small Small	Small Small	c
Mining	E M	Small Zero	Very small Very small	Small Very small	Small Very small	b

E = Effort
M = Market
a = Could be a large market for each of a given system
b = Systems which could have a high technical content
c = Could be relatively simple systems

REFERENCES

Chapter 1

1. *Proceedings of the Fluid Amplification Symposium*, Harry Diamond Laboratories, October 1962 Washington, May 1964 Washington, October 1965 Washington. Distributed by U.S. Department of Commerce, Clearinghouse for Federal Scientific and Technical Information.
2. *Proceedings of the Cranfield fluidics conferences*, British Hydromechanics Research Association, 1st Conference September 1965 Cranfield, 2nd Conference January 1967 Cambridge, 3rd Conference May 1968 Turin, 4th Conference March 1970 Coventry.
3. Bain, D. C., and Baker, P. J.: *Technical and market survey of fluidics in the United Kingdom*, British Hydromechanics Research Association, 1969.
4. Bourque, C., and Newman, B. G.: 'Reattachment of a two-dimensional incompressible jet to a flat plate', *Aero. Quart.*, 1960, Vol. 9, p. 201.
5. Sawyer, R. A.: 'Flow due to a two-dimensional jet issuing parallel to a flat plate', *Journal of Fluid Mechanics*, 1960, Vol. 9, p. 543.
6. Sawyer, R. A.: 'Two-dimensional reattaching jet flows including the effects of curvature on entrainment', *Journal of Fluid Mechanics*, 1963, Vol. 17, p. 481.
7. Jones, N. S., Foster, K., and Mitchell, D. G.: 'A method of calculating the wall pressure distribution in a turbulent reattachment bubble', *Proc. 2nd Cranfield Fluidics Conf.*, 1967, Paper A3.
8. Boucher, R. F.: 'Incompressible jet reattachment analysis using a good free jet model', *Proc. 3rd Cranfield Fluidics Conf.*, 1968, Vol. 1, Paper F1.
9. Kelly, L. R., and Shinn, J. N.: 'Noise in fluidic proportional amplifiers', *Proc. 3rd Cranfield Fluidics Conf.*, 1968, Vol. 2, Paper G1.
10. Beatty, E. K., and Markland, E.: 'Feasibility study of laminar jet deflection fluidic elements', *Proc. 3rd Cranfield Fluidics Conf.*, 1968, Vol. 2, Paper H1.
11. Verhelst, H. A. M.: 'On the design, characteristics and production of turbulence amplifiers', *Proc. 2nd Cranfield Fluidics Conf.*, 1967, Paper F2.

12. Glaettli, H. H., Müller, H. R., and Zingg, R. H.: 'Remarks on the limitations of pure fluid elements', *Proc. Fluid Amplification Symposium*, 1964, Vol. 1.
13. Rechten, A. W., and Zückler, T.: 'New aspects of miniaturization of fluid logic elements', *Proc. 3rd Cranfield Fluidics Conf.*, 1968, Paper F5.
14. Comparin, R. A., Mitchell, A. E., Müller, H. R., and Glaettli, H. M.: 'On the limitations and special effects in fluid jet amplifiers', Symposium on fluid jet control devices, *A.S.M.E.* Winter Ann. Meeting, Nov. 1962, p. 65.
15. Rechten, A. W.: 'Flow stability in bistable fluid elements', *Proc. 2nd Cranfield Fluidics Conf.*, 1967, Paper B6.
16. Syred, N., Royle, J. K., and Tippetts, J. R.: 'Optimisation of high gain vortex devices', *Proc. 3rd Cranfield Fluidics Conf.*, 1968, Paper J3.
17. Bahr, J.: 'The foil element, a new fluid logic element', *Elektronische Rechenanlagen*, 1965, vol. 7, no. 2, p. 69.

Chapter 2

1. Auger, R. N.: 'Turbulence amplifier design and applications', *Proc. Fluid Amplification Symposium*, Vol. 1, Harry Diamond Laboratories, Vol. 1, p. 357, 1962.
2. Verhelst, H. A. M.: 'On the design, characteristics and production of turbulence amplifiers', Paper F2, *2nd Cranfield Fluidics Conference*, Cambridge, January 1967.
3. Fox, H. L., and Goldschmied, F. R.: 'Basic requirements for an analytical approach to pure fluid control systems', *Proc. Fluid Amplification Symposium*, Harry Diamond Laboratories, Vol. 1, p. 293, 1964.
4. Goldschmied, F. R.: 'Fluid digital amplifiers of axi-symmetric focused-jet design', *Fluidics*, p. 215, Fluid Amplifier Associates, 1965.
5. Lechner, T. J., and Sorensen, P. H.: 'Some properties of the direct and transverse impact modulators', *Proc. Fluid Amplification Symposium*, Harry Diamond Laboratories, Vol. II, p. 35, 1964.
6. Bjornsen, B. J.: 'The impact modulator', *Proc. Fluid Amplification Symposium*, Harry Diamond Laboratories, Vol. II, p. 5, 1964.

Chapter 3

1. Kolff, *et al.*: *T. Amer. Soc. Art. Int. Org.*, Vol. 5, p. 298, 1959.
2. Akutsu, *et al.*: *Amer. Heart J.*, Vol. 59, p. 731, 1960.

3. Kolff, *et al.*: *T. Amer. Soc. Art. Int. Org.*, Vol. 8, p. 135, 1962.
4. Kantrowitz: *T. Amer. Soc. Art. Int. Org.*, Vol. 6, p. 292, 1960.
5. Woodward, *et al.*: 'A fluid amplifier artificial heart pump', *Harry Diamond Laboratory Symposium*, Washington, U.S.A. 1962, p. 81.
6. Brychta, O.: 'A logic system using a microdiaphragm device and its use in medical process control', *Proc. I.F.A.C.*, Paper C5, 1968, London.
7. Tsuchiya, K. *et al.*: *Japan Soc. mech. engr.*, No. 519, Vol. 73, 1967.
8. Tsuchiya, K. *et al.*: 'Artificial heart controlled by fluid amplifier', *Proc. I.F.A.C.*, Paper C7, 1968, London.
9. Woodward, Mon, Joyce, Straub and Barila: 'Four fluid amplifier controlled medical devices', *Harry Diamond Laboratory Symposium*, Washington, U.S.A., Vol. 4, 1964, p. 172.
10. Humphrey and Tarunoto: *Fluidics*, Fluid Amplifier Associates Inc., Boston, Mass., U.S.A., 1965, p. 169.
11. Joyce, Woodward and Barila: *Proc. 18th Annual Conference, Engineering in Medicine and Biology*, Phil. Pa., 1965, Vol. 7, p. 20.
12. Ball, G. J.: 'Fluidic control of automatic lung ventilators', M.Phil. thesis, University of London, 1968.
13. Straub, H. H.: 'Design requirements and proposal for an army respirator', Report No T.R.-1249, Harry Diamond Laboratories, Washington, U.S.A., 1964.
14. Straub, H. H., and Meyer, J.: 'An evaluation of a fluid amplifier face mask respirator', *Harry Diamond Laboratory Symposium*, Washington, U.S.A., 1965, Vol. 3, p. 309.
15. Pavlin, C., and Kadosch, M.: 'Mechanical characteristics of a pure fluid respirator with curved walls', *1st Cranfield Fluidics Conference*, 1965, pp. F3-33.
16. Vince and Brown: 'The application of fluid jet devices to a medical respirator', *1st Cranfield Fluidics Conference*, 1965, pp. F4-45.
17. Brown, C. C.: 'Fluid amplifier entrainment oscillator', Patent Application No. 30630/66.
18. Esmond, W. G., and Clark, H.: 'Mathematical analysis and mass transfer optimization of a compact, low cost, pumpless system for haemodialysis', *Biomedical Fluid Mechanics Symposium*, A.S.M.E., 1966, p. 161.
19. Montgomerie, G. A., and Floyd, T. J.: 'Heavy current fluidic devices', *Proc. 2nd Cranfield Fluidics Conference*, 1967, pp. K4-41.
20. Petrovick, M. L.: 'On the nature of fluid amplifiers and the oral cavity', *Trans. I.S.A.*, Vol. 7, p. 130, 1968.

Chapter 5

1. Grant, J., and Wright, J.: 'Potential applications of fluidics in nuclear plant', Paper K2, *Proc. 2nd Cranfield Fluidics Conference*, British Hydromechanics Research Association, 1967, pp. K2-9 to K2-19.
2. Fisher, M. J., and Thomson, A.: 'The application of fluidics to the detection of burst fuel elements in nuclear reactors', Paper K5, *Proc. 2nd Cranfield Fluidics Conference*, British Hydromechanics Research Association, 1967, pp. K5-53 to K5-58.
3. Halbach, C. R., Otsap, B. A., and Thomas, R. A.: 'A pressure insensitive fluidic temperature sensor', *Advances in fluidics*', American Society of Mechanical Engineers, 1967, pp. 298–312.
4. Schwartz, H. W.: *Liquid metal vapour quality meter*, Bendix Research Laboratories, Report No 3540, 60 pp.
5. Tesla, N.: *Valvular conduit*, U.S. Patent No 1,329,559. 3 February 1920.
6. Heim, R.: 'An investigation of the Thoma counter flow brake', *Trans. Munich Hydraulic Institute*, Bulletin No 3.
7. Baker, P. J.: 'A comparison of fluid diodes', Paper D6, *Proc. 2nd Cranfield Fluidics Conference*, British Hydromechanics Research Association, 1967, pp. D6-88 to D6-115.
8. Jacobs, B. E. A., and Baker, P. J.: 'The cascade diode', Paper K5, *Proc. 3rd Cranfield Fluidics Conference*, British Hydromechanics Research Association, 1968, pp. K5-63 to K5-82.
9. McArthur, G. L.: *Analytical study of the feasibility of a pneumatic neutron flux detector*, National Aeronautics and Space Administration, Report CR 54468, 1965, 94 pp.
10. Milleman, S. E.: *Development of a bread board model of an all-pneumatic neutron flux detector*, National Aeronautics and Space Administration, Report CR-72422, 1968, 180 pp.
11. Galli de Paratesi, S., Ghiurghi, L., and Musik, H.: 'Development of fluidic systems for nuclear plant control', Paper B6, *Proc. 3rd Cranfield Fluidics Conference*, British Hydromechanics Research Association, 1968, pp. B6-77 to B6-88.

Chapter 6

1. Leathers, J. W., Savage, A. C., and Pearson, I. P.: 'A fluid control system for diesel generators', Paper B2, *Proc. 3rd Cranfield Fluidics Conference*, British Hydromechanics Research Association, Cranfield, 1968, pp. B2-17 to B2-32.
2. Smith, W. L.: *Development of fluid logic start-up system for main forced draft blowers*, Naval Ship Engineering Center, Philadelphia, Report No AD 659499, 1967, p. 7.
3. Shamin, V. M.: 'Automation of marine diesel controls using

pneumatic logic elements (relays)', *Sudostronie*, Vol. 30, No 9, September 1964. (In Russian. English translation available from British Ship Research Association as Trans. No. 2050.)
4. Law, A. C.: 'Pneumonic control of marine engines', *Fluid Power International*, Vol. 33, p. 382, January 1968, pp. 40–44.
5. Law, A. C.: 'Fluidic control of marine diesel engines', *Fluidics International*, Vol. 1, 2, December 1968, pp. 24–29.
6. Dexter, E. M., and Colston, J. R.: 'Pure fluid controls for a ship's boilers', *Control Engineering*, Vol. 11, No 9, September 1964, pp. 96–97.
7. Reason, J.: 'Boiler control proves analog fluidics', *Control Engineering*, Vol. 14, No 3, March 1967, pp. 59–63.
8. Banham, J. W.: *A review of the state-of-the-art of fluidic control*, Naval Ship Engineering Centre, Philadelphia, Report No AD 667597, 1968, p. 19.
9. Humphrey, E. F., and Tarumoto, D. H.: *Fluidics*, Fluid Amplifier Associates, 1968, pp. 231–239.
10. Bowles, R. E., and Sowers, E. H.: 'A pure fluid hydrofoil control system', *Proc. Fluid Amplification Symposium*, Harry Diamond Laboratories, Vol. 2, AD 601,500, May 1964, pp. 431–458.
11. Colston, J. R., and Dexter, E. M.: 'Applications of Pure Fluid Techniques to a Speed Control', *Proc. Fluid Amplification Symposium*, Harry Diamond Laboratories, Vol. 2, AD 601,500, May 1964, pp. 405–430.
12. Boothe, W. A.: 'Feasibility study of a fluid amplifier steam turbine speed control', *Proc. Fluid Amplification Symposium*, Harry Diamond Laboratories, Vol. 2, AD 601,500, May 1964, pp. 377–404.
13. Anon.: 'Fluidic thruster steers a barge', *Machine Design*, Vol. 40, No 12, 23 May 1968, p. 16.
14. Dexter, E. M.: 'Applications to solutions of control problems', *Fluidics Quarterly*, Vol. 1, No 2, January 1968, pp. 103–110.
15. Buck, J. R.: 'Pure fluidic control system for submarine hover during missile launching', *Naval Engineers Journal*, October 1968, pp. 759–766.
16. Goldschmied, F. R.: *Hydraulic axisymmetric focused jet diverters with pneumatic control*, N.A.S.A. Technical Memorandum X-53554, December 1966, p. 27.
17. Goldschmied, F. R.: 'Underwater hovering control with fluidic amplifiers', *Journal of Hydronautics*, Vol. 2, No. 2, April 1968, pp. 102–107.
18. Bain, D. C., and Baker, P. J.: *Technical and market survey of fluidics in the United Kingdom*, British Hydromechanics Research Association, 1969, p. 35.

19. Buck, J. R.: 'Fluidic control systems applied to the control of buoyant pontoons', *Naval Engineers Journal*, Vol. 79, No. 6, December 1967, pp. 977–990.

Chapter 7

1. Clayton, B. J., and Posingies, W. M.: 'Development and flight test of a flueric roll control system', *Journal of Spacecraft and Rockets*, Vol. 4, No 2, Feb. 1967.
2. Eastman, N.: 'Missile control by fluidics', Paper C3, *Third Cranfield Fluidics Conference*, 1968.
3. Rivard, J. G.: 'Secondary injection t.v.c. using fluidic vortex valves', *Wescon '67*, Part 4.
4. Platt, T.: *Final report of hot gas t.v.s. research to February 1968*, B.A.C. Report No. R48/36/TVC/002/GEN (Restricted).
5. Rodgers, D. L.: *Development and flight testing of a fluidic flight control system*, N.A.S.A. CR-913.
6. Berg, D. F. *et al.*: 'Fluidics has a future in jet engine control', *Control Engineering*, July 1966.
7. Rose, R. K., and Phipps, W. L.: 'Fluidic control of a J79 turbo-jet engine', *Proc. A.S.M.E.* 67-WA/FE-33.
8. Rimmer, R.: 'A fluidic pressure ratio sensor for gas turbine engine control', Paper C3, *Proc. I.E.A.C.*, 1968.
9. Davies, G. E.: 'A fluidic method for inlet guide vane control for jet engines', Paper C6, *Proc. I.E.A.C.*, 1968.
10. Chadwick, V. J.: 'A method using wall-attachment devices in an ultra-sensitive mode', Paper B1, *2nd Cranfield Fluidics Conference*, January 1967.
11. Boothe, W. A., and Woodson, C. W.: 'A.C. fluidic systems', G.E. Report No 66-C-434.
12. Evans, R. A.: *Applied fluidic technology*, Honeywell Inc.

Chapter 8

1. Rolland, J. N.: 'Digital step pipe amplifiers application in gas distribution control', *3rd Cranfield Fluidics Conference*, 1968, Paper B7.
2. Small, D. A.: 'Optimizing fluidic element size for speed and power', *A.S.M.E. Winter General Meeting*, Pittsburgh, 1967, Paper 67-WA/FE-39.
3. Sarpkaya, T., and Kirshner, J. M.: 'The comparative performance characteristics of vented and unvented cusped, straight and curved-walled bistable amplifiers', *3rd Cranfield Fluidics Conference*, 1968, Paper F3.
4. Bidgood, R. E., Ramanathan, S., and Powell, A. E.: 'The future; fluidics in industry', Nottingham, 1968, Session 5, Paper 1.

5. Jensen, D. F., Mueller, H. R., and Schaffer, R. R.: 'Miniature diaphragm element: cheap, fast, predictable', *Control Engineering*, Nov. 1969, pp. 70–71.
6. Foster, K., *et al.*: 'A digital point-to-point positioning system', *3rd Cranfield Fluidics Conference*, Turin, 1968, Paper A1.
7. Kelly, L. R., and Shinn, J. N.: 'Noise in fluidic proportional amplifiers', *3rd Cranfield Fluidics Conference*, Turin, 1968, Paper G1.
8. Bowles, R. E., *et al.*: 'Fluidic a.c. amplifiers', *3rd Cranfield Fluidics Conference*, Turin, 1968, Paper G2.
9. Eastman, N.: 'Missile Control by Fluidics', *3rd Cranfield Fluidics Conference*, Turin, 1968, Paper C3.
10. Khol, R.: 'A.C. fluidics', *Machine Design*, Vol. 41, Feb. 1969, pp. 126–130.
11. Colston, J. R., and Dexter, E. M.: 'Applications of pure fluid techniques to speed control', *Proc. Fluid Amplification Symposium*, Washington, 1964, Vol. 11, pp. 377–404.
12. Colston, J. R.: 'A pure fluid turbine speed control', *A.S.M.E. Winter General Meeting*, New York, 1964, Paper 64-WA/AUT-8.
13. Boothe, W. A.: 'Feasibility study of a fluid amplifier steam turbine speed control', *Proc. Fluid Amplification Symposium*, Washington, 1964, Vol. 11, pp. 405–430.
14. General Electric Co.: *Fluid amplifier symbols, nomenclature and specification*, N.A.S.A. Report CR-147, 1965.
15. Tafel, H. J., and Kohn, A.: 'European work on fluidics standards', *Fluidics Feedback*, Vol. 3, No 7, July 1969, pp. 181–188.
16. Taft, C. K.: 'Hydraulic fluidics', *S.A.E. Farm, Construction and Industrial Machinery Meeting*, Milwaukee, 1967, Paper 670736.
17. Winslow, W. M.: 'Induced fibration of suspensions', *Journal of Applied Physics*, Vol. 20, Dec. 1949, pp. 1137–1140.
18. Johnson, E. G.: *Fluidic gas turbine controls*, Honeywell Publication M497.
19. Binder, A. M.: 'Throttle is only moving part', *S.A.E. Journal*, Vol. 75, No 8, Aug. 1967, pp. 36–38.
20. Humphrey, E. F., and Taramoto, D. H., ed., *Fluidics*, Fluidic Amplifier Associates Inc., Boston, 1963.
21. Bjornson, B. G.: 'A fluidic amplifier pneumatic controller', *Control Engineering*, June 1965, pp. 88–93.
22. Hass, H.: 'New approach to d.d.c. uses fluidics', *Control Engineering*, Feb. 1969, pp. 74–77.
23. Anon.: 'Fluidics controls yarn tension instrument', *Control Engineering*, Nov. 1968.
24. Bain, D. C., and Baker, P. J.: *Technical and market survey of fluidics in the United Kingdom*, B.H.R.A., Cranfield, 1969.

THE AUTHORS

E. Markland (*Chapter 1, Fundamentals of fluidics*) is Professor of Mechanical Engineering at University College, Cardiff. After graduating from the University of Manchester in 1944 his first appointment was in the Aerodynamics Department of the Royal Aircraft Establishment, Farnborough. In 1946 he joined the office of a London consulting engineer, and in 1949 moved to the University of Nottingham, where he developed extensive laboratories for fluid mechanics and thermodynamics. In 1966 he was elected to a Chair in the Mechanical Engineering Department of Queen's University, Belfast. Professor Markland moved to his present professorship in 1970. He has worked in fluidics since 1964.

R. F. Boucher (*Chapter 1, Fundamentals of fluidics*) is a lecturer on mechanical engineering at the University of Sheffield. He was a student apprentice and graduate trainee with the Central Electricity Generating Board in London. In 1962 he began his researches in fluidics at the University of Nottingham and became an I.C.I. Fellow there in 1966. Dr Boucher later moved to Queen's University, Belfast, and was appointed to a lectureship in 1968.

Anthony McCabe (*Chapter 2, Devices and their manufacture*) lectures on aeronautical engineering at Stockport College of Technology. Dr McCabe graduated in aeronautical engineering at the University of Manchester and obtained a Ph.D. in 1963 for research on supersonic aerodynamics. After a spell at Hawker-Siddeley Aviation as a specialist aerodynamicist he took a temporary lectureship in the Department of Fluid Mechanics, University of Manchester, in 1964. In 1966 he joined the English Electric Co., Whetstone, Leicester, now part of G.E.C. Power Engineering, and worked on fluidic device design and industrial applications.

Raymond Hazard (*Chapter 2, Devices and their manufacture*) is the Head of Structures and Installation in the Missile Department of B.A.C. He was trained as an instrument maker at Thomas Mercer Ltd and worked as an instrument fitter during his two years' service in the R.A.F. He was then trained as a design engineer at Elliott Bros (London) Ltd, where he obtained his H.N.C. in mechanical engineering. In 1953 he joined English Electric on guided weapon work, on which he is still employed, although the company is now the British Aircraft Corporation. At B.A.C. he has designed and developed control equipment utilizing hydraulics, pneumatics and electronics for both missiles, trackers and launchers. Before appointment to his present position he spent six years in more basic research on high-temperature hydraulic equipment and in developing the design of fluidic components and their integration as compact control units. He also assisted Geoffrey Gant in the engineering of the fluidic missile (Chapter 7).

G. A. Parker (*Chapter 3, Medical applications*) is Reader in Mechanical Engineering at the University of Surrey. He was educated at King Edward VI School, Nuneaton, and the University of Birmingham, graduating in 1957 with an honours degree in mechanical engineering. At Hawker-Siddeley Dynamics he worked on guided-weapon power-supply control systems until 1960, when he returned to Birmingham University. There his research on hydraulic control systems led to a Ph.D. Since then Dr Parker has worked on advanced machine-tool control systems at the Cincinnati Milling Machine Co. in the U.S.A. and he has been a lecturer at Birmingham University. He is now active in fluid power research, and has been particularly associated with fluidic control system development. Currently he is particularly interested in the control problems connected with biomedical engineering.

G. M. Brewin (*Chapter 4, Industrial applications*) is a senior executive with Mullard, Southampton. On leaving Loughborough College School he became a student apprentice with the British United Shoe Machinery Co. of Leicester. There followed two years' National Service, spent servicing and maintaining radar and associated video systems in the R.A.F. After periods in the research divisions of both the British United Shoe Machinery Co. and Brush Electrical Engineering Co. he joined the English Glass Co. of Leicester in 1961. In 1965 Brewin moved to Cambridge to lead the newly formed Fluidics Division of Techne (Cambridge) Ltd. He was responsible for all aspects of the company's activities in this field, including the presentation of several papers to international conferences. He is the originator of various patents, ranging from road-marking devices to automatic control equipment.

M. J. Fisher (*Chapter 5, Nuclear applications; Chapter 6, Marine applications*) is an Assistant Director of Research at the British Hydromechanics Research Association. His early experience included some years in the metal industry and in the chemical industry. He then moved into the missile industry and spent five years on control system design. In 1960 he joined the B.H.R.A., at first working on seals. His present interests are fluidics, fluid power and instrumentation.

G. C. Gant (*Chapter 7, Aerospace applications*) is the Head of the Mechanical Technology Department, B.A.C. Guided Weapons Division, Stevenage. He started his career with a seven-year student apprenticeship to Davey Paxman & Co, Colchester. During his apprenticeship he obtained his H.N.D. in mechanical engineering at the North East Essex Technical College. Then he went up to King's College, Durham University (now University of Newcastle-upon-Tyne), and graduated with first-class honours in mechanical engineering. After graduating he joined Regulateurs Europa (a totally owned subsidiary of Davey Paxman) and worked on engine/generator-set control equipment for two years. His next move was to the Control Department of B.A.C.'s Guided Weapons Division, Stevenage, where he began as a research engineer on special hydraulic accumulator development. A year later he was a section leader in charge of fluidic autopilot and high-speed solenoid research projects. Three years after that, in May 1969, he was appointed to his present post. The work of his department covers research in control engineering, fluid dynamics and structures, specialist support to the product design department and main project stressing.

B. J. Cooper (*Chapter 8, Trends and prospects*) is Head of the Control Group in the Mechanical Engineering Laboratory of the English Electric Co. He served a sandwich course apprenticeship at Brush Electrical Engineering and took an external engineering degree in 1952. From 1954 to 1956 he was with G.E.C., working on guided-weapon electronic systems. After that he joined the Mechanical Engineering Laboratory of the English Electric Co. and worked on all types of industrial control system. He was appointed Head of the Control Group in 1964.

D. L. Hughes (*Chapter 8, Trends and prospects*) is Head of the Control Section of the English Electric Co.'s Mechanical Engineering Laboratory. He served an apprenticeship at the company's Rugby works before going to the University College of Swansea to read for a degree in mechanical engineering. After graduating he stayed four years at the university as a member of the Engineering Department staff. He rejoined English Electric in 1966.

Arthur Conway (*Editor*) is a chartered engineer. He served a five-year apprenticeship to the de Havilland Aircraft Co. and took his degree in mechanical engineering at Imperial College. His experience has ranged from the light end (aircraft industry) to the heavy end (iron and steel industry), including a short spell in nuclear power. He has been the Editor of the technical journals *Control*, *Instrument Review* and *The Engineer*, and he writes for both the general and the technical press.

INDEX